Deep Eutectic Solvents in the Textile Industry

Amjad Farooq · Hafeezullah Memon ·
Aamir Farooq · Zongqian Wang

Deep Eutectic Solvents in the Textile Industry

Springer

Amjad Farooq
School of Textile and Garment
Anhui Polytechnic University
Wuhu, China

School of Textile and Clothing
Qingdao University
Qingdao, China

Aamir Farooq
School of Automation
Nanjing University of Science
and Technology
Nanjing, China

Hafeezullah Memon
College of Textile Science
and Engineering
International Institute of Silk
Zhejiang Sci-Tech University
Hangzhou, China

Zongqian Wang
School of Textile and Garment
Anhui Polytechnic University
Wuhu, China

ISBN 978-981-97-6432-7 ISBN 978-981-97-6433-4 (eBook)
https://doi.org/10.1007/978-981-97-6433-4

This Springer imprint is published by the registered company Springer Nature Singapore Pte Ltd.
The registered company address is: 152 Beach Road, #21-01/04 Gateway East, Singapore 189721, Singapore

Preface

Over the past few decades, the development of nanocellulose (NC) from various lignocellulosic sources has emerged as a prominent area of research in the field of sustainable chemistry and engineering. NC possesses intriguing properties such as sustainability, biodegradability, high surface area, exceptional mechanical strength, and low toxicity, positioning itself as a promising nature-based material for numerous applications including nanocomposites, packaging materials, electronics, and more. In recent years, NC has also garnered increasing attention as a thickener and stabilizer for Pickering emulsions, attributed to its high surface area and flexibility. The exceptional properties of nanocellulose make it a highly promising material with significant potential for diverse applications in various fields. These applications include 3D printing and biomedicine, food packaging, functional nano papers, Pickering emulsions, flexible electronics, and wearable sensors. In recent years, there has been a notable increase in research efforts focused on the preparation of nanocellulose, specifically cellulose nanofibers and cellulose nanocrystals. Cellulose nanofibrils are considered materials that have specific dimensions with a diameter of less than 100nm and longer such as several micrometers. Subsequently, cellulose nanocrystals are needle-shaped nanofibers having dimensions such as a diameter of less than 20 nm and lengths varying from 50 to 500 nm.

Traditionally, the production of NC involves the removal of lignin from lignocellulosic materials, which is often a costly and labor-intensive process, contributing to environmental pollution. Moreover, the surface of NC contains numerous hydrophilic groups, necessitating hydrophobic modifications to tailor its properties for Pickering emulsions. However, these modifications not only increase production costs but also pose potential safety risks. Typically, such modifications involve multi-step chemical treatments and the consumption of various chemicals, which can lead to environmental pollution. Thus, scientists are trying to find ways to extract nanocellulose by using green and environmentally friendly methods in which deep eutectic solvent (DES) pretreatment for the production of nanocellulose has been widely considered due to its favorable properties. DESs can be easily prepared using cost-effective components and offer high tunability. DES system possesses the capability to perform a range of structural and chemical functionalization techniques on cellulosic fibers. Notably, they allow for esterification, succinylation, cationization, and sulfation, expanding the potential applications of these fibers. Additionally, DESs can be recycled through relatively simple methods and reused, further adding to their appeal as a sustainable option for nanocellulose production.

Deep eutectic solvents, first named by Abbott, belong to a class of solvents similar to ionic liquids and share similar characteristics such as high thermal stability, low vapor pressure, non-combustibility, and tunability. Notably, DESs possess additional advantages such as non-toxicity, biodegradability, affordability, and ease of synthesis. Generally, DESs have lower molecular weights compared to ionic liquids and are formed through intermolecular hydrogen bonding between hydrogen bond donors (HBDs) and hydrogen bond acceptors (HBAs). DESs are composed of two or more pure compounds, and their melting points are typically lower than those of the individual components. This study provides an in-depth exploration of the importance of deep eutectic solvents (DESs) in the production of nanocellulose, regardless of the cellulose sources utilized. These sources encompass a wide range of materials such as wheat straw, rice straw, corn stalk, sugarcane bagasse, rapeseed stem, grape pomace, elephant grass, and various others. The development of nanocellulose from various lignocellulosic sources, coupled with the utilization of deep eutectic solvents, has emerged as a promising avenue for advancing sustainable chemistry and engineering in the textile industry. In recent years, nanocellulose has gained attention for its role in functional textiles, offering improved mechanical strength, moisture management, UV protection, and antibacterial activity. By incorporating nanocellulose into textile fibers or coatings using DESs, the textile industry can enhance the functionality and sustainability of textile materials, reducing reliance on synthetic additives and promoting environmentally friendly production processes. Additionally, the integration of nanocellulose enables the development of smart textiles with sensing capabilities and the fabrication of nanofiber membranes for advanced filtration applications. Thus, the use of nanocellulose, facilitated by DESs, holds great promise for revolutionizing the textile industry by providing sustainable, functional, and smart textile solutions. This book comprehensively discusses the significance of DESs in enabling nanocellulose production from diverse cellulose feedstocks and their valuable textile applications.

Wuhu/Qingdao, China — Amjad Farooq
Hangzhou, China — Hafeezullah Memon
Nanjing, China — Aamir Farooq
Wuhu, China — Zongqian Wang

About This Book

Nanocellulose has been considered widely in various applications because of its exceptional intrinsic characterization including high aspect ratio, excellent biocompatibility, easy surface modification, and adjustable optical properties. These unique and excellent characteristics make nanocellulose highly versatile for a wide range of applications. However, nanocellulose produced by conventional methods possesses challenges in terms of economics and environmental impact, hindering them from being prepared for large-scale production. Consequently, researchers are trying to explore alternative approaches, and one emerging option is the utilization of deep eutectic solvents (DESs) for nanocellulose preparation. Thus, the excellent biodegradable and biocompatible nature of deep eutectic solvents, coupled with their low toxic features, easy formation, impressive tunable characteristics, and recyclability, make them highly outstanding substitutes for the preparation of nanocellulose. In this study, we explore recent advancements in the extraction of nanocellulose using deep eutectic solvents, which are considered promising solvents for separating biopolymers. The study specifically focuses on identifying significant properties of cellulose and deep eutectic solvents. It also highlights the utilization of deep eutectic solvents in purifying the cellulose for possible preparation of cellulose nanocrystals (CNCs) and cellulose nanofibrils (CNFs). The results obtained so far are promising regarding the utilization of DESs for nanocellulose production as the favorable recyclability and reusability of DESs suggest that they hold great potential as viable options for large-scale production of nanocellulose and nanocellulose-based products. By connecting these exceptional properties of deep eutectic solvents and their compatibility with cellulose, this book paves the way for sustainable and efficient methods of producing nanocellulose, enabling its broader application in the textile industry and beyond.

Contents

About the Authors

Dr. Amjad Farooq earned his Bachelor of Science degree in Textile Engineering from the College of Textile, Bahauddin Zakariya University, Multan, Pakistan, in 2012. Following his undergraduate studies, he gained valuable industry experience as a Senior Officer of Quality Control at Masood Textile Mills in Faisalabad, Pakistan, from 2012 to 2014. In 2017, Dr. Farooq completed his Master of Science degree at the College of Textile Engineering, Donghua University, Shanghai, China. His dedication and outstanding performance during his master's studies earned him the prestigious International Student Award. Continuing his academic journey, he successfully received his Ph.D. degree from the College of Textile, Donghua University, Shanghai, in 2022. During his doctoral and master's studies, Dr. Farooq's research focused on the utilization of green chemicals for the extraction and application of nanocellulose derived from agricultural waste.

In recognition of his exceptional research and academic achievements, Dr. Farooq was honored with the Excellent Ph.D. Graduate award from Donghua University in 2022. He has published seven research articles and has made significant contributions to the field. Currently, Dr. Farooq serves as a Postdoctoral Fellow at Anhui Polytechnic University, Anhui, China, where he continues his research on the utilization of nanocellulosic materials obtained from renewable resources. His expertise lies in textile fibers and materials applications, with a particular focus on the development of nanomaterials. Dr. Farooq's research has resulted in the publication of nearly 50 papers, showcasing his dedication to advancing knowledge in his field.

Through his extensive academic background, industry experience, and numerous publications, Dr. Amjad Farooq has established himself as a specialist and scholar in the field of textile fibers and materials applications. His commitment to sustainable and innovative research contributes to the development of nanomaterials and their applications, paving the way for advancements in the field of nanocellulose-based materials.

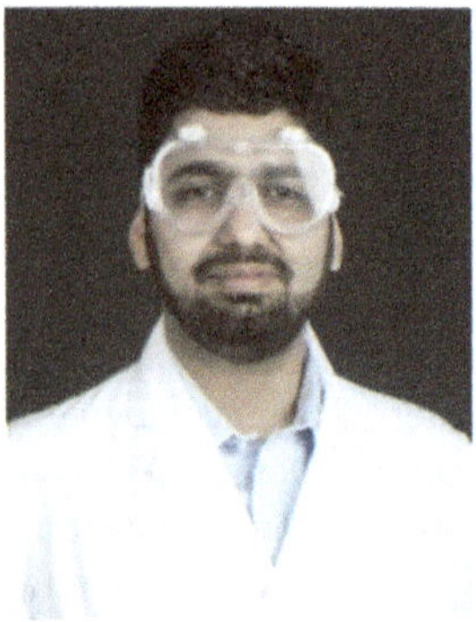

Dr. Hafeezullah Memon received his B.E. in Textile Engineering from Mehran University of Engineering and Technology, Jamshoro, Pakistan, in 2012. He served at Sapphire Textile Mills as Assistant Spinning Manager for more than one year while earning his master's in Business Administration from the University of Sindh, Pakistan. He completed his master's in Textile Science and Engineering from Zhejiang Sci-Tech University, China, and a Ph.D. in Textile Engineering from Donghua University in 2016 and 2020. Dr. Memon focuses on fiber-reinforced composites, textiles and management, and biobased materials. Since 2014, Dr. Memon has published more than 60 peer-reviewed technical papers in international journals and conferences, filed more than ten patents, edited more than ten books and special issues, and worked on more than ten industrial projects. Dr. Memon is a Fellow of the Textile Institute, a full professional, and an active member of the Society for the Advancement of Material and Process Engineering, the Textile Society of America, the American Chemical Society, the Society of Wood Science and Technology, and the International Textile and Apparel Association. Moreover, he is a registered Engineer of the Pakistan Engineering Council. He has served as a reviewer and editorial board member of several international journals and has reviewed more than 600 papers. Dr. Memon is a recipient of several national and international awards. Currently, he is serving as Foreign Talent Expert, Distinguished Associate Professor, after completing his postdoc at Zhejiang Sci-Tech University. Moreover, he is elected as Chairman of the Alumni Association of International Students at Zhejiang Sci-Tech University.

Mr. Aamir Farooq received his Bachelor of Science degree in Electrical Engineering from the Federal Urdu University of Arts Science and Technology, Pakistan, in 2015. In 2018, He received his Master of Science degree from the College of Electrical Engineering, Shandong University, China. He has researched Power Electronics and Power Transmission systems. At present, Aamir is a doctorate degree student at Nanjing University of Science and Technology, Nanjing, China. His research interests are AC-DC converter, buck-boost converter, and control strategy of Vienna rectifier. Aamir is actively involved in the research and development of new energy technology and control system engineering. He has published more than eight papers and some academic works.

Dr. Zongqian Wang earned his Ph.D. in Textile Science and Engineering. In 2018, he was appointed as a visiting researcher at the University of California, Davis, USA. Following the completion of his doctoral studies, Dr. Wang joined Anhui Polytechnic University, where he has been actively engaged in scientific research focusing on innovative functional materials for textiles. In 2019, he was appointed as a professor and assumed the role of a doctoral supervisor.

Dr. Wang holds several noteworthy positions, including Director of the Anhui Ecological Textile Printing and Dyeing Manufacturing Innovation Center, Leader of the Applied Peak Discipline of Textile Science and Engineering in Anhui Province, and Chief Scientist of the Anhui High-Performance Materials Innovation Center. He has successfully led over 20 scientific research projects, encompassing prestigious funding sources such as the National Natural Science Foundation of China, Anhui Provincial Major Science and Technology Projects, Anhui Provincial Key Research and Development Program, and Anhui Provincial Natural Science Foundation's Collaborative Innovation Project with Anhui Provincial Universities. Additionally, Dr. Wang has undertaken more than 40 research and development projects commissioned by various enterprises.

Dr. Wang's exceptional leadership and expertise have been recognized through numerous accolades. One of his research projects received the first prize for scientific and technological progress in Anhui Province. In 2023, he was selected as an innovative A-class talent in the "Special Support Plan" of Anhui Province, further solidifying his reputation as a prominent figure in academia and research. Moreover, Dr. Wang has been acknowledged as an academic and technical leader in Anhui Province. In 2020, he was recognized as a top talent in young and middle-aged disciplines (majors) among universities in Anhui Province.

Chapter 1
Identification of Cellulose and Nanocellulose

Abstract Nanocellulose has emerged as a promising nanomaterial with numerous applications in various scientific and industrial domains. This chapter provides an overview of different types of nanocellulose, namely cellulose nanofibrils (CNFs), cellulose nanocrystals (CNCs), and bacterial cellulose (BC). CNFs are long and flexible nanofibers with high aspect ratios, obtained through mechanical or enzymatic processes. They exhibit exceptional mechanical properties, biodegradability, and high surface area, making them suitable for applications in reinforced composites, barrier films, and nanocomposites. CNCs, on the other hand, are rod-like nanoparticles with high crystallinity and aspect ratios, obtained through acid hydrolysis of cellulose fibers. CNCs possess unique optical, mechanical, and rheological properties, enabling their utilization in various fields such as biomedical engineering, electronics, and packaging materials. Bacterial cellulose is produced by certain bacteria, notably Gluconacetobacter xylinus, through the fermentation of glucose. It forms a three-dimensional network of nanofibers with high purity and unique properties, including high water-holding capacity, biocompatibility, and mechanical strength. This chapter highlights the distinct characteristics of CNFs and CNCs. Their unique physicochemical properties make them highly valuable for a wide range of applications in diverse fields. Understanding the different types of nanocellulose opens up opportunities for the development of innovative and sustainable materials with enhanced properties and performance.

1.1 Introduction

With an expected annual production of 1.5×10^{12} tons, cellulose has been considered so abundant on Earth that it is thought to be an unlimited source of raw materials. The fact that it is a renewable, biodegradable, and non-toxic substance makes it a viable solution for addressing the present environmental issues. The structural and chemical functionalization of cellulosic fibers enables their use as fillers, packaging films, textiles, paper goods, building materials, and many other diverse applications. Naturally, cellulose fibers are found in various sources such as wood, plants, animals,

A. Farooq et al., *Deep Eutectic Solvents in the Textile Industry*,
https://doi.org/10.1007/978-981-97-6433-4_1

algae, fungi, and bacteria. However, wood and cotton are recognized as the primary sources of this biopolymer [1, 2]. Cotton contains nearly pure cellulose, but cellulose in wood is coupled to hemicelluloses and both are linked to lignin by hydrogen and covalent bonding, indicating to hire of several procedures to extract cellulose (e.g., Kraft process). Cellulose can be efficiently derived through two distinct approaches. The top-down approach involves utilizing wood, cotton, annual plants, or agricultural residues to obtain cellulose of specific sizes. On the other hand, the bottom-up approach entails biosynthesizing cellulose from glucose through bacterial processes [3, 4]. Naturally, cellulose materials exist primarily in the shape of elementary fibrils that are subsequently formed into bigger units named micro/nanofibrils, eventually, they are aggregated into cellulosic fibers (Fig. 1.1). In evaluating the dimensions, generally, the diameter of elementary fibrils is believed to be between 3.5 to 5 nm, while the diameter of nanofibrils spans from 10 to 50 nm. However, the lengths of nanofibrils may range from hundreds of nanometers to a few microns [5]. Mechanical processes can be employed to disintegrate fibril aggregates in two ways. Firstly, they can disintegrate the aggregates, resulting in the production of cellulose nanofibrils (CNF). Alternatively, they can divide the aggregates transversally at the less ordered regions, leading to the formation of cellulose nanocrystals [6].

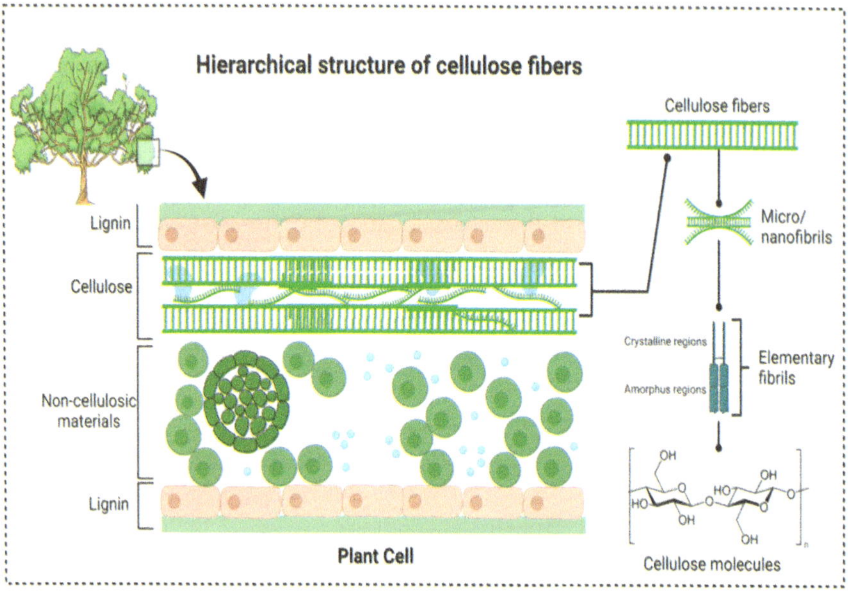

Fig. 1.1 Hierarchical structural illustration of cellulose fibers

1.2 Nanocellulose

Nanocellulose is a fibrous material extracted from cellulose with at least dimensional parameters as a nanometric scale having lengths from 1 to 100 nm. They possess excellently interesting characteristics like broader specific surface area from 100 to 200 g/m^2, tremendous mechanical properties of prepared films such as stiffness up to 30 GPa and tensile strength of about 300 MPa, and easier to employ surface modification process and can alter their aspect ratios [7, 8]. Thus, these distinct features enable them to have a wide range of applications, including renewable energy devices, food packaging films, biodegradable nanocomposite materials, biomedicine, and so on [9]. Generally, nanocelluloses have been classified into three types: (1) cellulose nanofibrils, which are also known as nanofibrillated cellulose, microfibrillated cellulose, or cellulose nanofibers, (2) cellulose nanocrystals, which are also known as nanocrystalline cellulose or cellulose whiskers, and (3) bacterial nanocellulose (BNC). According to the reports, in contrast to CNF and CNC, the latter has been prepared by the bottom-up method, however, BNC is generally formed by fermentation of low molecular weight carbohydrates by bacteria mostly by using Gluconacetobacter xylinum [9, 10].

1.2.1 Cellulose Nanocrystals

Cellulose nanocrystals with extensive domains of crystalline regions have been named nano-whiskers, nanorods, and rod-like cellulose crystals [11–13] as well, they are produced by a biosynthesis process from natural fiber's cellulose [14, 15]. Cellulose nanocrystals are regarded as broadly rigid as compared to cellulose nanofibrils and are commonly obtained from cellulose fibers by acid hydrolysis process. Generally, cellulose nanocrystals are obtained by pretreatment of natural fiber biomass with alkali and bleaching treatment to eliminate noncellulosic constituents. Later on, acid hydrolysis is performed with specific reaction conditions including temperature, agitation, and at the end washing of the residue with ionized water by using centrifugation techniques to remove excessive chemicals. Later on, freeze-drying or spray-drying methods are used to prepare pure cellulose nanocrystals. From the previous reports, it has been mentioned that the reaction conditions including the nature of acid and the source of raw material for cellulose extraction have great effects on the dimensions and morphology of cellulose nanocrystals [16]. In acid hydrolysis commonly sulfuric acid (64 wt%) is used to obtain CNC, and cellulose nanowhiskers exhibit relative width of 3–50 nm and length has been evaluated between 100 nm to several micrometers [17]. Several reports mentioned that the cellulose nanocrystals obtained from various sources have exhibited excellent characteristics including low density (1.5–1.6 g/cm^3), excellent thermal stability (260 °C), high aspect ratio (10–70), higher crystallinity (54–88%) and cellulose yield (100%) [17–20], various types of nanocellulose and their extraction techniques is shown in Fig. 1.2.

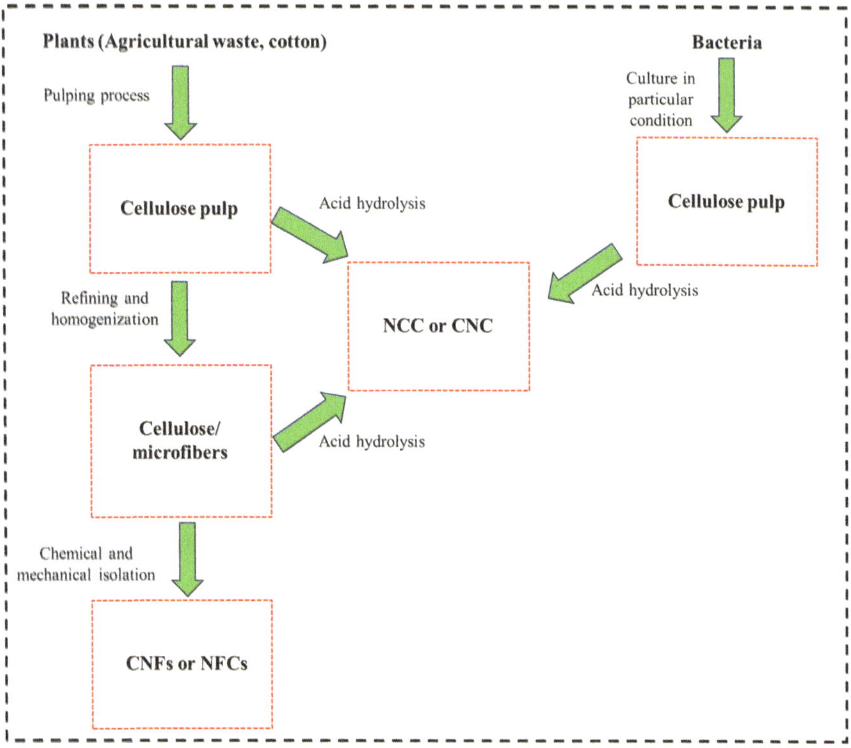

Fig. 1.2 Various types of nanocellulose and their extraction techniques

1.2.2 Cellulose Nanofibrils

Cellulose nanofibrils were introduced by Turbak and coworkers in 1983 [21] who obtained cellulose with horizontal dimensions (in nanometers), Table 1.1 shows more details. They passed the softwood pulp aqueous solution at various times through a high-pressure homogenizer. However, during this treatment, the obtained residue was nanofibers containing both amorphous and crystalline regions because of high shearing forces. Obtained nanomaterials exhibited a higher aspect ratio and when exposed to an aqueous solution presented gel form in water. Cellulose fibers from natural raw materials can be allowed to form cellulose nanofibrils with excellent morphology and dimensions depending on the processing conditions used in this whole process. Generally, CNFs show dimensions as a diameter of about 5 to 50 nm and a few micrometers in length [22]. Cellulose nanofibrils have been evaluated as a natural cellulosic constituent containing a higher percentage of cellulose with specific dimensions such as larger volume, high surface area, lower diameter (20 to 60 nm), and length in several micrometers.

Table 1.1 Dimensional characteristics of cellulose nanofibrils and cellulose nanocrystals obtained from various sources

Types	Sources	Width (nm)	Length (nm)	References
Cellulose nanocrystals	Plant	3–30	100–500	[15, 23, 24]
	Wood	3–6	100–200	[25, 26]
	Bacteria	10–50	100–1000	[27, 28]
	Tunicate	5–30	100–2000	[29, 30]
	Algae	5–30	Higher than 1000	[31, 32]
Cellulose nanofibrils	Plant	3–100	Higher than 1000	[33, 34]
	Wood	2–100	Higher than 1000	[35, 36]
	Bacteria	20–100	Different dimensions	[26, 37]
	Tunicate	10–30	Higher than 2000	[38–40]

1.2.3 Bacterial Cellulose

Bacterial cellulose (BC) is a form of cellulose produced by bacteria in aqueous culture media containing a sugar source. The production of BC typically takes several days to two weeks. BC has the same chemical composition as cellulose derived from plants, but it lacks functional groups like carboxyl or carbonyl, which are commonly introduced during the purification process of plant cellulose. Additionally, BC is free of other polymers such as lignin, hemicelluloses, or pectin, making it ideal for biomedical applications. BC is synthesized as twisting ribbons with cross-sections measuring approximately 3–4 nm wide and 70–140 nm long. The length of BC ribbons can exceed 2 μm. It has a high degree of polymerization (DP) ranging from 3000 to 9000, indicating a large molecular weight, and exhibits a crystallinity level of 80–90%, which is notably high [41]. The unique characteristics of BC, including its high purity, absence of impurities, and well-defined nanostructure, make it highly suitable for a range of biomedical applications. Its high crystallinity and mechanical strength make it a valuable scaffold material for tissue engineering. Its biocompatibility and ability to retain significant amounts of water make it advantageous for wound dressing applications. Furthermore, BC's nanoscale structure and large surface area offer opportunities for further functionalization and modification to enhance its properties and expand its application potential [42].

1.3 Conclusion

Cellulose, as the most abundant and renewable organic matter on Earth, represents an immense resource estimated at approximately 1 trillion tons. Its significance lies in its role as a source of natural fibers, making it a primary feedstock for the production of a wide range of materials. These materials include, but are not limited to, paper, board, textiles, powders, and reinforcing fillers. The versatility of cellulose as a raw material enables its utilization across various industries, contributing to the development of diverse products and applications. Cellulose nanofibrils and cellulose nanocrystals are being prepared from cellulose for a wide range of required applications. Bacterial cellulose produced by certain microorganisms has gained significant recognition as a multifunctional nano-biomaterial. Similar to plant cellulose, BC is composed of linear glucan molecules that are interconnected through hydrogen bonds. This structural similarity to plant cellulose contributes to the unique properties and functionalities exhibited by BC.

References

1. Heinze, T. 2016. Cellulose: Structure and properties. In *Cellulose chemistry and properties: Fibers, nanocelluloses and advanced materials*, 1–52.
2. Klemm, D., et al. 2005. Cellulose: Fascinating biopolymer and sustainable raw material. *Angewandte Chemie International Edition* 44 (22): 3358–3393.
3. Mateo, S., et al. 2021. Nanocellulose from agricultural wastes: Products and applications—A review. *Processes* 9 (9): 1594.
4. Dufresne, A., M.A.S. Azizi Samir, F. Alloin. 2005. Review of recent research into cellulosic whiskers, their properties and their application in nanocomposite field. *Biomacromolecules* 6: 612–626.
5. Lavoine, N., et al. 2012. Microfibrillated cellulose–Its barrier properties and applications in cellulosic materials: A review. *Carbohydrate Polymers* 90 (2): 735–764.
6. Habibi, Y. 2014. Key advances in the chemical modification of nanocelluloses. *Chemical Society Reviews* 43 (5): 1519–1542.
7. Fang, Z., et al. 2019. Nanocellulose-based films and their emerging applications. *Current Opinion in Solid State and Materials Science* 23 (4): 100764.
8. Phanthong, P., et al. 2018. Nanocellulose: Extraction and application. *Carbon Resources Conversion* 1 (1): 32–43.
9. Pradhan, D., A.K. Jaiswal, and S. Jaiswal. 2022. Emerging technologies for the production of nanocellulose from lignocellulosic biomass. *Carbohydrate Polymers*, 119258.
10. Abitbol, T., et al. 2016. Nanocellulose, a tiny fiber with huge applications. *Current opinion in Biotechnology* 39: 76–88.
11. Dujardin, E., M. Blaseby, and S. Mann. 2003. Synthesis of mesoporous silica by sol–gel mineralisation of cellulose nanorod nematic suspensions. *Journal of Materials Chemistry* 13 (4): 696–699.
12. John, M.J., and S. Thomas. 2008. Biofibres and biocomposites. *Carbohydrate Polymers* 71 (3): 343–364.
13. Iwamoto, S., A. Nakagaito, and H. Yano. 2007. Nano-fibrillation of pulp fibers for the processing of transparent nanocomposites. *Applied Physics A* 89: 461–466.
14. Kumar, R., et al. 2019. A simple approach for the isolation of cellulose nanofibers from banana fibers. *Materials Research Express* 6 (10): 105601.
15. Garcia de Rodriguez, N.L., W. Thielemans, and A. Dufresne. 2006. Sisal cellulose whiskers reinforced polyvinyl acetate nanocomposites. *Cellulose* 13: 261–270.
16. Habibi, Y., L.A. Lucia, and O.J. Rojas. 2010. Cellulose nanocrystals: Chemistry, self-assembly, and applications. *Chemical Reviews* 110 (6): 3479–3500.
17. Hubbe, M.A., et al. 2008. Cellulosic nanocomposites: A review. *BioResources* 3 (3): 929–980.

18. Jonas, R., and L.F. Farah. 1998. Production and application of microbial cellulose. *Polymer Degradation and Stability* 59 (1–3): 101–106.
19. Dufresne, A. 2010. Processing of polymer nanocomposites reinforced with polysaccharide nanocrystals. *Molecules* 15 (6): 4111–4128.
20. Dufresne, A. 2006. Comparing the mechanical properties of high performances polymer nanocomposites from biological sources. *Journal of Nanoscience and Nanotechnology* 6 (2): 322–330.
21. Turbak, A.F., F.W. Snyder, and K.R. Sandberg. 1983. Microfibrillated cellulose, a new cellulose product: Properties, uses, and commercial potential. In *Journal of applied polymer science: Applied polymer symposium*
22. Nechyporchuk, O., M.N. Belgacem, and J. Bras. 2016. Production of cellulose nanofibrils: A review of recent advances. *Industrial Crops and Products* 93: 2–25.
23. Habibi, Y., et al. 2008. Bionanocomposites based on poly (ε-caprolactone)-grafted cellulose nanocrystals by ring-opening polymerization. *Journal of Materials Chemistry* 18 (41): 5002–5010.
24. Li, Q., J. Zhou, and L. Zhang. 2009. Structure and properties of the nanocomposite films of chitosan reinforced with cellulose whiskers. *Journal of Polymer Science Part B: Polymer Physics* 47 (11): 1069–1077.
25. Klemm, D., et al. 2011. Nanocelluloses: A new family of nature-based materials. *Angewandte Chemie International Edition* 50 (24): 5438–5466.
26. Peciulyte, A., et al. 2015. Impact of the supramolecular structure of cellulose on the efficiency of enzymatic hydrolysis. *Biotechnology for Biofuels* 8 (1): 1–13.
27. Araki, J., and S. Kuga. 2001. Effect of trace electrolyte on liquid crystal type of cellulose microcrystals. *Langmuir* 17 (15): 4493–4496.
28. Roman, M., and W.T. Winter. 2004. Effect of sulfate groups from sulfuric acid hydrolysis on the thermal degradation behavior of bacterial cellulose. *Biomacromolecules* 5 (5): 1671–1677.
29. Habibi, Y., H. Chanzy, and M.R. Vignon. 2006. TEMPO-mediated surface oxidation of cellulose whiskers. *Cellulose* 13: 679–687.
30. Jalal Uddin, A., J. Araki, and Y. Gotoh. 2011. Extremely oriented tunicin whiskers in poly (vinyl alcohol) nanocomposites. *Polymer International* 60 (8): 1230–1239.
31. Moon, R.J., et al. 2011. Cellulose nanomaterials review: Structure, properties and nanocomposites. *Chemical Society Reviews* 40 (7): 3941–3994.
32. Revol, J.-F. 1982. On the cross-sectional shape of cellulose crystallites in Valonia ventricosa. *Carbohydrate Polymers* 2 (2): 123–134.
33. Chinga-Carrasco, G. 2011. Cellulose fibres, nanofibrils and microfibrils: The morphological sequence of MFC components from a plant physiology and fibre technology point of view. *Nanoscale Research Letters* 6: 1–7.
34. Zambrano, F., et al. 2020. Using micro-and nanofibrillated cellulose as a means to reduce weight of paper products: A review. *BioResources* 15 (2): 4553–4590.
35. Dufresne, A. 2017. Cellulose nanomaterial reinforced polymer nanocomposites. *Current Opinion in Colloid & Interface Science* 29: 1–8.
36. Gumrah Dumanli, A. 2017. Nanocellulose and its composites for biomedical applications. *Current Medicinal Chemistry* 24 (5): 512–528.
37. Klemm, D., et al. 2018. Nanocellulose as a natural source for groundbreaking applications in materials science: Today's state. *Materials Today* 21 (7): 720–748.
38. Zhao, Y., and J. Li. 2014. Excellent chemical and material cellulose from tunicates: Diversity in cellulose production yield and chemical and morphological structures from different tunicate species. *Cellulose* 21: 3427–3441.
39. Zhao, Y., et al. 2015. Tunicate cellulose nanocrystals: Preparation, neat films and nanocomposite films with glucomannans. *Carbohydrate Polymers* 117: 286–296.

40. Zhao, Y., et al. 2017. Cellulose nanofibers from softwood, hardwood, and tunicate: Preparation–structure–film performance interrelation. *ACS Applied Materials & Interfaces* 9 (15): 13508–13519.
41. Lahiri, D., et al. 2021. Bacterial cellulose: Production, characterization, and application as antimicrobial agent. *International Journal of Molecular Sciences* 22 (23): 12984.
42. Kumar, M., V. Kumar, and S. Saran. 2024. Current research and advances in the molecular aspect of bacterial cellulose synthesis with special emphasis on the production of bacterial cellulose from waste and its food applications. *Cellulose*, 1–29.

Chapter 2
Nanocellulose Isolation Techniques

Abstract Nanocellulose extraction has been of great interest to researchers over the years. Nanocellulose, derived from cellulose from different sources, has obtained significant attention because of its exceptional mechanical, optical, and barrier properties. This chapter provides an overview of lignocellulosic biomass for nanocellulose isolation, along with a discussion of common extraction methods and general applications. The focus of this chapter is primarily on description and explanation of nanocellulose extraction techniques which are commonly used in recent research. Additionally, a forward-looking perspective on the future of nanocellulose is presented. The aim is to offer insights and guidance for the efficient extraction of nanocellulose.

2.1 Introduction

Wood and plants are natural cellular hierarchical biocomposite materials produced by nature, in which semicrystalline cellulose microfibril-reinforced amorphous matrices consist of lignin, hemicellulose, waxes, and extractives [1]. Thus, lignocellulose fibers consist of microfibril collectives. Additionally, a plant's structure consists of multiple length scales, allowing for efficient use of materials while maintaining high strength. For example, wood consists of nearly 40 to 50 wt% cellulose in which half of the region is covered with nanocrystalline and the other half is in amorphous form as shown in Fig. 2.1. The whole tree length is in meters, the cross-sectional structure is described in centimeters, while growth rings are millimeters, cellular anatomy is exhibited in tens of micrometers, while the micrometers describe the layer structure within cell walls, tens of nanometers are exhibiting the configuration of cellulose fibrils containing the matrix consists on hemicellulose and lignin, and molecular structure of cellulose, hemicellulose, lignin, and their chemical interactions describes in nanometers [2].

Demand for utilization of the products prepared from sustainable materials has increased over the years. Cellulose is the most abundant biopolymer on Earth, which is biodegradable, renewable, cheap, and nontoxic. Cellulose separation from the

A. Farooq et al., *Deep Eutectic Solvents in the Textile Industry*,
https://doi.org/10.1007/978-981-97-6433-4_2

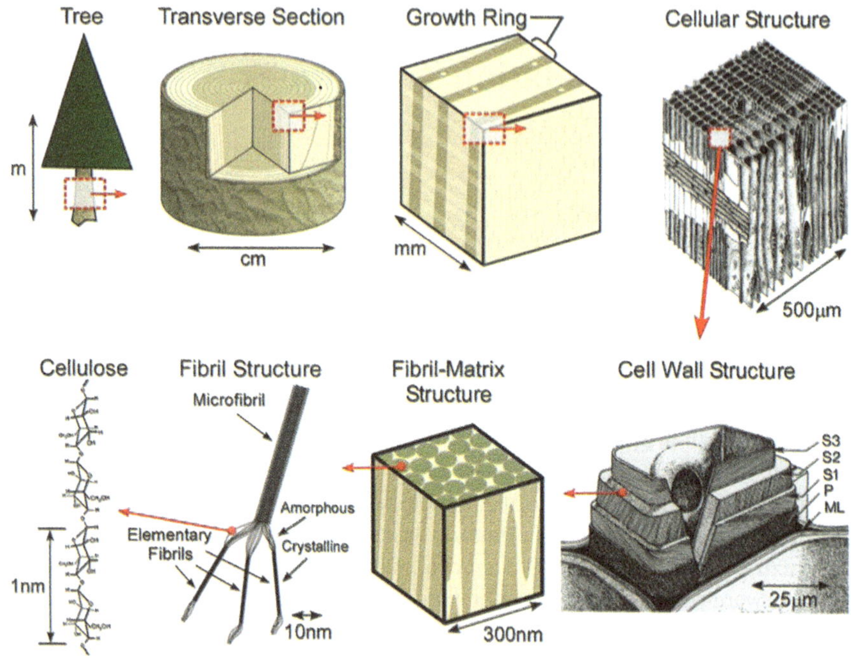

Fig. 2.1 Wood hierarchical structure: from tree to cellulose [3]. Reproduced with permission

plant fibers is being involved through various techniques conventionally by chemical treatments consisting of alkali treatment and bleaching. Because of their excellent hierarchical composition and semicrystalline nature, nanocellulose is extracted from this naturally occurring polymer by using a mechanically or chemically induced deconstructing technique. Cellulose nanoparticles or nanocellulose potential is being considered widely for special functional nanomaterials [4] and as a bio-based reinforcing nanofiller, these nanoparticles have attracted significant attraction over the last few years [5–7].

Over the last few decades people have been aware of environmental conservation, thus applications of renewable resources are widely studied and applied in their daily lives because they are friendly. Nanocellulose materials are naturally biodegradable and can be extracted from plant cell walls. Nanocellulose has been reported excellent material because of its excellent properties including high strength, high surface area values, and nanometer size in diameter [3, 8]. Additionally, nanocellulose consists of a higher amount of hydroxyl groups which help in surface modification. Nanocellulosic materials are used in different applications in our daily life applications including biomedical products, cosmetics, nanocomposite materials, and pharmaceutical applications. Recently, versatile applications of nanocellulose have attracted high attention of researchers and industrialists. Thus, this chapter focuses on generally used techniques for cellulose nanofibrils and cellulose nanocrystal extraction.

2.2 Lignocellulosic Biomass

Lignocellulosic biomass comprises a diverse range of natural organic materials, primarily derived from plants. They represent the largest and most promising group of sustainable carbon materials, serving as an excellent feedstock for the production of biochemicals, bioethanol, and biofuels [9]. Interestingly, lignocellulosic biomass is a valuable source of natural fibers that possess environmentally friendly properties, making them a potential substitute for petroleum-based polymers. Moreover, agricultural waste derived from biomass offers significant potential for reuse as fuel or feedstock, enabling the production of high-value materials without competing with human and animal food supplies. The cell wall structure of lignocellulosic biomass consists of majorly three types of polymers: lignin, hemicellulose, and cellulose. However, the composition and content of these components vary depending on the species, types, and sources of the lignocellulosic biomass [10, 11]. Lignin typically accounts for approximately 10–25% of the dry weight of lignocellulosic biomass [12–14]. In plant cell walls, lignin acts as a binder, providing structural integrity by binding cellulose and hemicellulose together and surrounding them. This binding function plays an important role in stiffness, compressive strength, resistance to deformation, and water impermeability of the plant cell wall. Moreover, lignin has been reported as a complex amorphous copolymer formed by the random combination of three phenylpropane monomers: p-coumaryl, coniferyl, and sinapyl alcohols. These monomers can have zero, one, or two methoxyl groups, respectively [15–17]. The ratio of these primary monomer units varies depending on the species and sources of lignocellulosic biomass. Recent research has focused on the isolation and depolymerization of lignin from lignocellulosic biomass for the production of biofuels and chemicals using natural materials [10, 16, 18]. Furthermore, lignin-based carbon materials have gained attention for their applications in catalysis, energy storage, and pollutant removal [17].

2.3 Cellulose Nanofibrils Extraction

2.3.1 *Mechanical Treatments*

Mechanical treatments are used to isolate cellulose nanofibrils from cellulose by using various techniques. Generally, high-pressure homogenization, crushing and grinding [19, 20], twin screw extrusion [21, 22], microfluidization [23, 24], ball milling [25, 26], high-intensity ultra-sonication [27, 28]. In the crushing and grinding process, the cellulose pulp is ground by passing it between a static stone and a rotating stone that rotates at around 1500 rpm. The shearing forces produced by the stones will shatter the cell wall structure, resulting in the individualization of the CNFs [29]. The relative distance between two stones, the shape of the stone channels, and the number of passes are all crucial factors in determining the dimensions of cellulose nanofibrils [30].

Whereas, in high-pressure homogenization, low-consistency cellulose pulp is generally pumped at an extensive pressure and subjected to pass through a series of valves that open and close in rapid sequence, putting the fibers to a considerable pressure drop as well as impact and shear stresses. Cellulose nanofibril's dimensions will be determined by the number of homogenizer passes and the amount of pressure applied [30]. The primary disadvantages of this specific technique are the utilization of high energy consumption and probable blockage issues that may emerge during the homogenization process [31]. Another promising mechanical treatment for cellulose nanofibrils extraction is twin screw extrusion. In this technique, two co-rotating screws have been set together in a closed barrel connection to fibrillate cellulose pulp. Finally, the cellulose suspension is subjected to heat, compression, and shear stresses as it goes through the extruder, resulting in fiber elimination [30, 32]. As an alternative to high-pressure homogenization, micro-fluidization was invented and applied for cellulose nanofibrils extraction. In this process, the cellulose suspension is driven under very high pressure by an intensifier pump into small Z- or Y-shaped slots and passes through them. Thus, by this process, the high shear as well as impact forces created by these slots will disrupt the hydrogen bonds, causing cellulose fibrillation. The dimensions of CNF are also determined by the micro-fluidizer's slot/chamber morphology as well as the number of cycles used in generating CNF to pass through the micro-fluidizer [33]. Another approach for isolating cellulose nanofibrils has been used is high-intensity ultra-sonication. In this process, high-intensity ultrasonic pulses (>20 kHz) are used and during the process these waves are absorbed by liquid molecules causing a mechanical oscillating motion, resulting in the production, development, and integration of gas bubbles which creates effective hydrodynamic pressures to obtain cellulose nanofibers [34]. Thus the obtained cellulose nanofibril's morphology is dependent on the power and duration of the ultra-sonication process [35]. Figure 2.2 shows the various mechanical treatments used in traditional CNF extraction methods.

2.3.2 Enzymatic Hydrolysis

To extract cellulose nanofibrils Enzymatic hydrolysis has also been used which is a biological treatment utilized to modify biomass cellulose fiber by decomposing the non-cellulosic impurities including lignin and hemicellulose. Generally, cellobiohydrolase, endoglucanase, and cellulase are enzymes employed in this pretreatment process which help to aid the restrictive hydrolysis of several particles or the selective hydrolysis of defined components in cellulose fibers [36]. As enzymatic treatments are complex, the enzyme action is based on the catalysis of the H-bonding linkage between the cellulose microfibers in the treatment process [37]. Since cellulosic fibers contain several organic compounds, so enzymatic treatment by a single

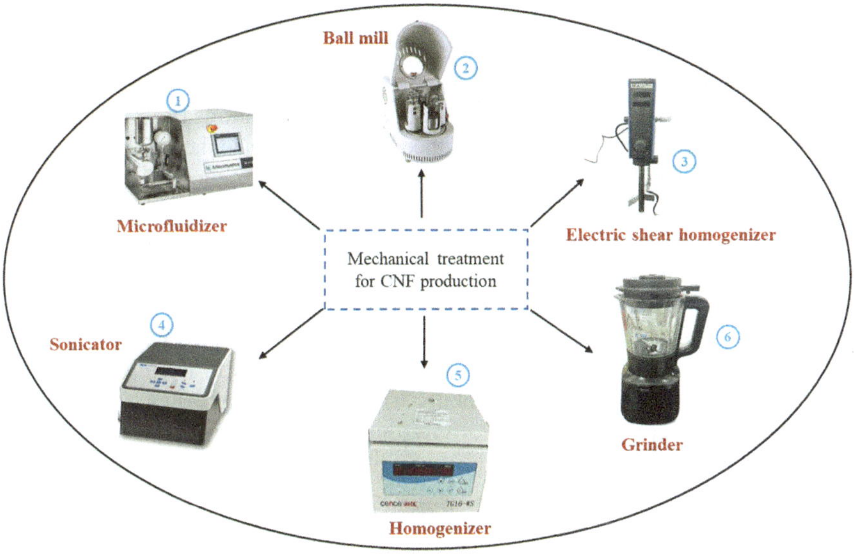

Fig. 2.2 Most applied mechanical treatment processes for cellulose nanofibril extraction

enzyme cannot degrade all non-cellulosic constituents, thus a set of selective cellulose enzymes have been used [38]. These selective enzymes have been concise as cellobiohydrolases: A- and B-type cellulases attack particularly crystalline cellulose, and endoglucanases: C-type and D-type cellulases attack cellulose fibers with disordered structures. There is a significant difference between the reaction times required for the enzymatic hydrolysis and acid hydrolysis of cellulose fiber, especially under mild conditions [36]. In comparison with acid hydrolysis, selective and mild hydrolysis using a mono-component endoglucanase enzyme was more effective in obtaining nanocellulose materials from softwood pulp, resulting in higher values of aspect ratio and less aggressive as compared to acid hydrolysis [39, 40].

2.3.3 *TEMPO Oxidation*

Cellulose nanofibrils have been prepared by using TEMPO-mediated oxidation which was primarily applied to polysaccharides for CNF preparation [23]. As a result, it has been commonly used for nanocellulose fibrillation. In this process, cellulosic fibers are carefully induced by the negatively charged carboxyl groups because of the electrostatic repulsions caused between negatively charged fibers [23, 24]. TEMPO oxidation pretreatment involves the selective oxidation of carboxylic groups by using sodium chlorite (NaClO) acting as a primary oxidant. In this process, catalytic percentage of 2,2,6,6-tetramethylpiperidine-1-oxyl radical (TEMPO) and sodium bromide (NaBr) is also present by maintaining pH in alkaline conditions between 10 to11 at constant room temperature [25]. Nevertheless, the TEMPO oxidation system has been considered to cause a higher degree of depolymerization of cellulose chains, due to the results of β-elimination of glycosidic bonds because of the occurrence of C6 aldehyde groups under alkaline conditions (Fig. 2.3) [26, 27]. Saito and his coworkers [27] have explained how to minimize the extensive depolymerization by using long reaction times and higher temperatures under natural or weak acidic pH of the solution under the TEMPO/NaClO/$NaClO_2$ system. In another study, Tanaka et al. explained that $NaClO_2$ used in this system is acting as a primary oxidant which is used to oxidize the formed aldehydes into carboxylic groups [28]. Obtained cellulose nanofibrils from mechanical treatments resulted in a high degree of fibrillation and energy saving, thus TEMPO oxidation process was resulted and scaled up to produce fibrillated cellulose, however, it has also been considered as high-cost process and un-environmentally friendly procedure. Thus, to reduce the cost of the reaction process Sanchez-Salvador [29] analyzed the possible way of reusing the reaction medium. In this study, two types of cellulose from different sources (cotton and eucalyptus) were used and 75% of the reaction media was used for ten cycles and fresh medium (25%) in each cycle was introduced.

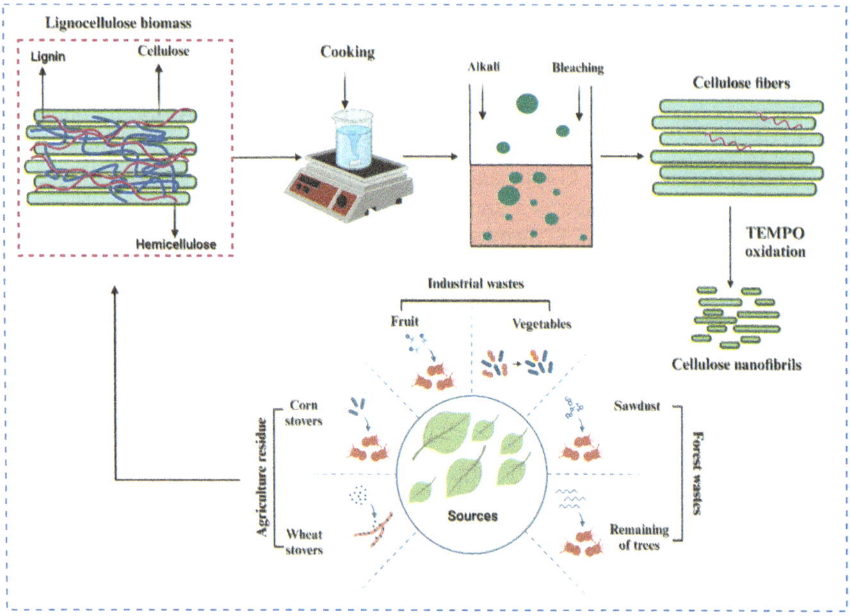

Fig. 2.3 Schematic illustration for cellulose nanofibrils extraction by using TEMPO oxidation technique

2.3.4 Cationization

Cellulose nanofibrils have also been prepared by using the cationization pretreatment technique for the first time by Aulin et al. [41]. Basically, in this method, the electrostatic repulsion between quaternary ammonium groups into cellulosic material structure, causing to increase in cellulose nanofabrication [42]. Various types of reactants have been evaluated and used in different studies to apply the cationization modification on cellulose structure such as 2,3-epoxypropyl trimethylammonium chloride, 2-hydrazinyl-2-oxoethyl) trimethylazanium chloride, GT, and chlorocholine chloride [41, 43, 44]. Following the cationization technique [45] uses oat straw pulp to extract cellulose nanofibrils. In this study, quaternary ammonium groups were allowed into cellulose fibers with chlorocholine chloride in the presence of dimethylsulfoxide (DMSO) and sodium hydroxide (NaOH), followed by microfluidization. Similarly, in the year 2014 Liimatainen et al. [44] evaluated tremendously a new way of cationization by utilizing periodate oxidation and subsequently used quaternization with Girard's reagent T. Later on high-pressure homogenization was done to extract cellulose nanofibrils and results from this study revealed that the dimensions of obtained CNF were excellent (10–50 nm in width) and cation charges were reported to appropriate as well (1.1–2.1 meq/g).

2.3.5 Pretreatment by Ionic Liquids

Ionic liquids pretreatment has been extensively used after alkali-acid and enzymatic pretreatments because of their excellent low vapor pressure, thermally stable and nonflammable as well as low melting point (less than 100 °C) properties [46–48]. Sugarcane bagasse cellulose fibers were dissolved by using ionic liquid 1-butyl-3-methylimidazolium chloride, later on, high-pressure homogenization was done to obtain cellulose nanofibrils. Results revealed that the cellulose fibers dissolution was dependent on the concentration of ionic liquid against cellulose fibers, total reaction time, temperature adopted for the system, and microwave power for this experiment [49].

2.3.6 Alkaline-Acid Pretreatment

Alkaline acid chemical pretreatment has been used for a long time for natural fibers generally this process is involved in removing hemicellulose, extractives, lignin, and waxes, in this process, cellulosic materials are treated with alkali such as sodium hydroxide/potassium hydroxide [50]. Alkaline pretreatments cause fibrillate and swelling of cellulosic fibers resulting in enhancing their surface area and ductility [51]. Additionally, natural fibers have been chemically modified with the help of

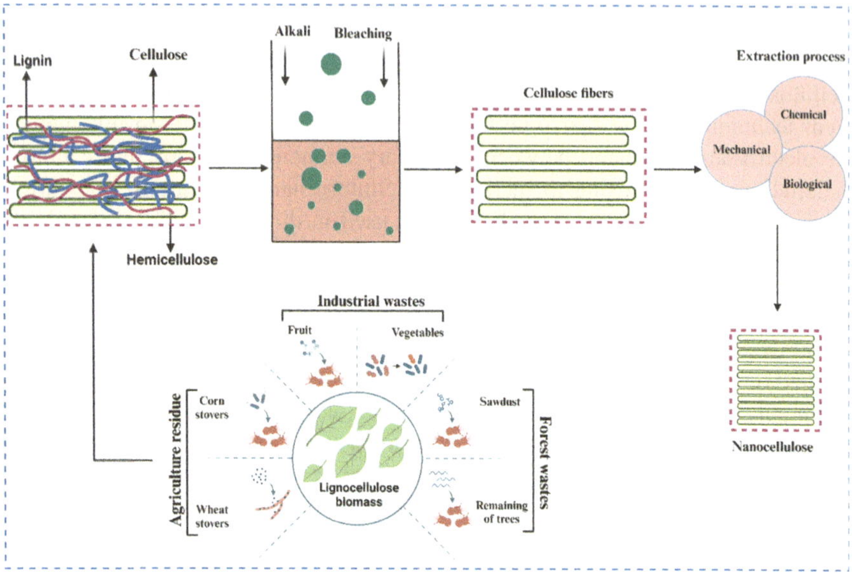

Fig. 2.4 Cellulose sources and various nanocellulose extraction techniques

silanes in which self-condensation is generally allowed with the siloxane towards OH-groups of cellulose fibers [52]. In various other reports, alkaline pretreatment has collaborated with the mechanical isolation of cellulose nanofibrils to eliminate non-cellulosic constituents (lignin, cellulose, and pectin) [53, 54]. Generally, alkaline pretreatment with mechanical disintegration has been carried out in three common steps such as; improving the surface area of cellulosic materials by hydrolysis with the help of pretreatment of natural fibers with sodium hydroxide solution 12–17.5 wt% for 2 h, subsequently, treatment of residue with a hydrochloric acid solution at a specific temperature (60–80 °C) to remove hemicellulose, and finally, remaining natural fiber residue treatment NaOH solution with a specific concentration (2 wt%) to eliminate remaining lignin [55]. Figure 2.4 shows the extraction process of cellulose nanofibrils.

2.4 Cellulose Nanocrystals Production

Cellulose nanocrystals have been evaluated as rod-like and needle-like cellulose crystals exhibiting excellent dimensions and morphological characteristics such as crystallinity up to 90%, diameter within the range of 5 to 70 nm, and length between 100 nm to several micrometers. The most common methods used in CNC extraction are acid hydrolysis, ionic liquid treatment, enzymatic pretreatment, and oxidation [56–63]. Additionally, like cellulose nanofibrils, CNC dimensions and properties are also dependent on their cellulosic sources, preparation methods, and chemical and experimental conditions [56, 64].

2.4.1 Acid Hydrolysis

Cellulose nanocrystals have been created in various ways however, acid hydrolysis has been extensively used in CNC isolation (Fig. 2.5). In acid hydrolysis, glycosidic bonds of disrupted domains of cellulose fibers are hydrolyzed by using mineral acids (mostly), however, crystalline domains are maintained and acid cannot be easily allowed to penetrate well-ordered crystalline domains [65]. After the hydrolysis process, the resultant suspension is carefully diluted in water suspension to stop the process. Later on, deionized water is used to eliminate the residual acid and other chemicals by using dialysis techniques. Deionized water is changed every three hours to remove the residual chemicals. Subsequently, the sonication process is carried out to get a homogenized dispersion of cellulose nanocrystals [56]. Generally specific concentration of hydrochloric acid and sulfuric acid is carried out for cellulose fiber hydrolysis, cellulose nanocrystals obtained by sulfuric acid show high colloidal stability because of the presence of charge groups (sulphate ester groups) on cellulose fiber surfaces during hydrolysis. However, CNC obtained by hydrochloric acid hydrolysis exhibits a coagulation tendency because of the absence of charged groups. To remove the environmental issues connected with the usage of these strong acids, formic, oxalic, and maleic acids have also been evaluated and used for cellulose nanocrystal preparation [66–68].

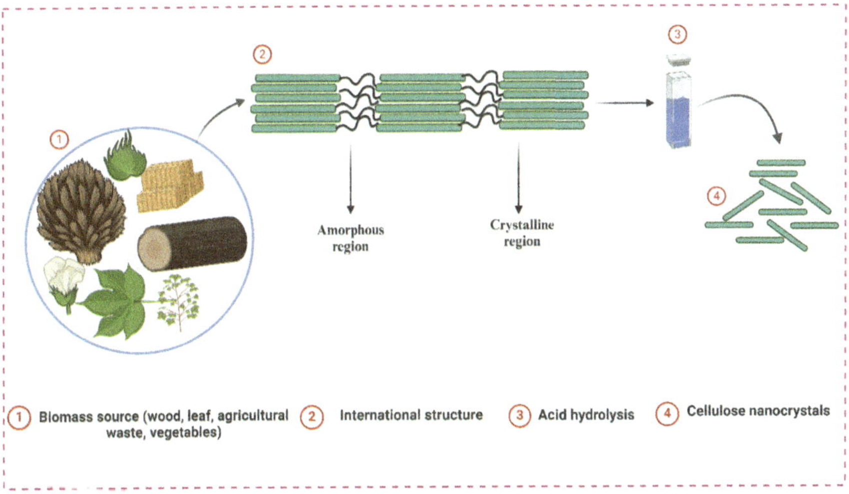

Fig. 2.5 A simple schematic illustration of acid hydrolysis technique for cellulose nanocrystals extraction

2.4.2 CNCs Preparation by Ionic Liquids

Ionic liquid materials named green solvents can be used for various special applications because they are easy to recover and can melt at low temperatures [69, 70]. Recently, ILs are applied in lignocellulosic materials handling [42, 71, 72]. Cellulose from cotton gin motes was applied for the production of cellulose nanocrystals by using ILs (dilute 1-butyl-3-methyl-1H-imidazol-3-ium chloride [BMIm][Cl]) and acids through a commonly practiced technique named as acidic method [71]. In this method, ionic liquids were applied to cellulose fibers and later on, hydrolysis of cellulose was done by sulfuric acid or hydrochloric acid. Jordan and co-workers in this report revealed the prepared CNC's crystallinity was comparatively the same as that prepared by traditional techniques. Additionally, the obtained cellulose nanocrystals show colloidal stability even more than 90 days proving an outstanding characteristic for CNC applications. In another report [73], cellulose nanocrystals were isolated through natural sources including cotton and eucalyptus (pulp) by using a homogenization technique in [BMIm][Cl] with higher pressure for various cycles (50). A similar method was applied to isolate CNC from bagasse by treating bagasse with ionic liquid and high-pressure homogenization was applied [48]. The results obtained from this study revealed that the physical treatments have greatly decreased the prepared CNC dimensions, especially its size. Zhao and co-workers [74] prepared CNCs through the nature-friendly and simple one-pot method by utilizing [BMIm][HSO4] as a catalyst/solvent and enzymatically esterification. Two different types of ILs (1-propyl-3-methylimidazol-3-ium chloride [PMIM][Cl] and 1-ethyl-3-methylimidazolium chloride [EMIM][Cl]) have been used in a study by Song et al. [75] to increase the thermal stability values of cellulose nanocrystals. Similarly, CNCs were prepared and compared by using two types of solvents [PMIM][Cl] and [EMIM][Cl], results revealed that the CNCs prepared by using [PMIM][Cl] were finer in dimensions (20 nm) as compared to [EMIM][Cl] (100 nm) [76]. Ionic liquids possess excellent advantages; however, their usage has been restricted recently because of the questions arising about their biodegradability, high cost, and toxicity in the isolation of cellulose nanocrystals.

2.4.3 CNC Preparation by Other Methods

CNC extraction has been done by using enzymatic hydrolysis and TEMPO-mediated oxidation, as in a recent study sugarcane bagasse cellulose was used to isolate CNC by enzymatic hydrolysis at high solid loading strategy [61]. The yield of cellulose nanocrystals in this study was about 50%, excellent crystallinity was about 96.5%, and dimensions in diameter were noted around 20 nm. While TEMPO oxidation is mostly used to extract cellulose nanofibrils, however, Cao et al. and coworkers have been able to create CNC at a higher yield (~80%), and excellent dimensions (diameter within 3 to 10 nm) were noted [62]. Although the above-mentioned protocols are being used to produce cellulose nanofibrils and cellulose nanocrystals, however, from a few years scientists and researchers have been focusing on the propagation of green and environmentally friendly methods to isolate nanocellulosic materials. Thus, an alternative way is getting tremendous attention to the use of deep eutectic solvents in pretreatment techniques against cellulose fibers. Thus, for an abundant and green production of cellulose and nanocellulose deep eutectic solvents have been used widely. The development of tailored pretreatment techniques to extract polymers from biomass has been considered extensively important. Pretreatment of lignocellulose biomass successfully by DES is reported to be dependent on its molar ratio, component types used in DES, and the type of biomass itself. Generally, the different biomass varies from each other because of growth conditions, structural versatility, and differences in composition lignin content [77].

2.5 Conclusions

Despite the excellent potential properties and abundant availability of nanocellulose in nature, the extraction process from lignocellulosic biomass or cellulosic materials remains a significant challenge. The strong aggregation of lignin, hemicelluloses, and other substances within the plant cell wall requires biomass pretreatment steps as a crucial step to eliminate non-cellulosic materials. However, such pretreatment methods often involve complex procedures, long periods, the use of toxic chemicals, and the production of wastewater. Moreover, conventional techniques for nanocellulose extraction from cellulosic materials have certain drawbacks. Acid hydrolysis, for instance, results in the generation of a substantial amount of acidic wastewater, while the mechanical process requires high energy consumption. Enzymatic hydrolysis, on the other hand, entails long reaction times. The motivation behind nanocellulose research lies in the development of novel extraction methods that are simple, produce minimal or no wastewater, and surpass the properties and yield achieved by traditional methods. Therefore, there is a need to explore innovative approaches for nanocellulose extraction that offer simplified steps, reduced wastewater generation, and improved properties and yield compared to conventional techniques. In which deep eutectic solvent can be a better choice.

References

1. Fengel, D., and G. Wegener. 2011. *Wood: Chemistry, ultrastructure, reactions*. Walter de Gruyter.
2. David, I. 1964. McGraw-Hill yearbook of science and technology. *Soil Science* 98 (4): 278.
3. Dufresne, A. 2013. Nanocellulose: A new ageless bionanomaterial. *Materials Today* 16 (6): 220–227.

4. Lin, N., J. Huang, and A. Dufresne. 2012. Preparation, properties and applications of polysaccharide nanocrystals in advanced functional nanomaterials: A review. *Nanoscale* 4 (11): 3274–3294.
5. Azizi Samir, M.A.S., F. Alloin, and A. Dufresne. 2005. Review of recent research into cellulosic whiskers, their properties and their application in nanocomposite field. *Biomacromolecules* 6 (2): 612–626.
6. Eichhorn, S.J., et al. 2010. Current international research into cellulose nanofibres and nanocomposites. *Journal of Materials Science* 45: 1–33.
7. Moon, R.J., et al. 2011. Cellulose nanomaterials review: Structure, properties and nanocomposites. *Chemical Society Reviews* 40 (7): 3941–3994.
8. Dufresne, A. 2012. Nanocellulose: Potential reinforcement in composites. *Natural Polymers* 2: 1–32.
9. Lee, H., S. Hamid, and S. Zain. 2014. Conversion of lignocellulosic biomass to nanocellulose: Structure and chemical process. *The Scientific World Journal* 2014.
10. Langan, P., et al. 2014. Common processes drive the thermochemical pretreatment of lignocellulosic biomass. *Green Chemistry* 16 (1): 63–68.
11. Agbor, V.B., et al. 2011. Biomass pretreatment: Fundamentals toward application. *Biotechnology Advances* 29 (6): 675–685.
12. Burhenne, L., et al. 2013. The effect of the biomass components lignin, cellulose and hemicellulose on TGA and fixed bed pyrolysis. *Journal of Analytical and Applied Pyrolysis* 101: 177–184.
13. Demirbaş, A. 2005. Estimating of structural composition of wood and non-wood biomass samples. *Energy Sources* 27 (8): 761–767.
14. Li, X., et al. 2015. Determination of hemicellulose, cellulose and lignin in moso bamboo by near infrared spectroscopy. *Scientific Reports* 5 (1): 17210.
15. Isikgor, F.H., and C.R. Becer. 2015. Lignocellulosic biomass: A sustainable platform for the production of bio-based chemicals and polymers. *Polymer Chemistry* 6 (25): 4497–4559.
16. Azadi, P., et al. 2013. Liquid fuels, hydrogen and chemicals from lignin: A critical review. *Renewable and Sustainable Energy Reviews* 21: 506–523.
17. Liu, W.-J., H. Jiang, and H.-Q. Yu. 2015. Thermochemical conversion of lignin to functional materials: A review and future directions. *Green Chemistry* 17 (11): 4888–4907.
18. Wang, H., M. Tucker, and Y. Ji. 2013. Recent development in chemical depolymerization of lignin: A review. *Journal of Applied Chemistry* 2013(9).
19. Rambabu, N., et al. 2016. Production of nanocellulose fibers from pinecone biomass: Evaluation and optimization of chemical and mechanical treatment conditions on mechanical properties of nanocellulose films. *Industrial Crops and Products* 83: 746–754.
20. Supian, M.A.F., et al. 2020. Production of cellulose nanofiber (CNF) from empty fruit bunch (EFB) via mechanical method. *Journal of Environmental Chemical Engineering* 8 (1): 103024.
21. Espinosa, E., et al. 2019. Production of lignocellulose nanofibers from wheat straw by different fibrillation methods. Comparison of its viability in cardboard recycling process. *Journal of Cleaner Production* 239: 118083.
22. Ho, T.T.T., et al. 2015. Nanofibrillation of pulp fibers by twin-screw extrusion. *Cellulose* 22: 421–433.
23. Lee, S.-Y., et al. 2009. Preparation of cellulose nanofibrils by high-pressure homogenizer and cellulose-based composite films. *Journal of Industrial and Engineering Chemistry* 15 (1): 50–55.
24. Taheri, H., and P. Samyn. 2016. Effect of homogenization (microfluidization) process parameters in mechanical production of micro-and nanofibrillated cellulose on its rheological and morphological properties. *Cellulose* 23 (2): 1221–1238.
25. Yu, W., et al. 2019. Choline chloride-based deep eutectic solvent systems as a pretreatment for nanofibrillation of ramie fibers. *Cellulose* 26: 3069–3082.
26. Zhang, L., T. Tsuzuki, and X. Wang. 2015. Preparation of cellulose nanofiber from softwood pulp by ball milling. *Cellulose* 22: 1729–1741.

27. Amiralian, N., et al. 2015. Isolation of cellulose nanofibrils from Triodia pungens via different mechanical methods. *Cellulose* 22: 2483–2498.
28. Frone, A.N., et al. 2011. Preparation and characterization of PVA composites with cellulose nanofibers obtained by ultrasonication. *BioResources* 6 (1): 487–512.
29. Siró, I., and D. Plackett. 2010. Microfibrillated cellulose and new nanocomposite materials: A review. *Cellulose* 17: 459–494.
30. Kargarzadeh, H., et al. 2018. Advances in cellulose nanomaterials. *Cellulose* 25: 2151–2189.
31. Djafari Petroudy, S.R., et al. 2021. Recent advances in cellulose nanofibers preparation through energy-efficient approaches: A review. *Energies* 14 (20): 6792.
32. Nechyporchuk, O., M.N. Belgacem, and J. Bras. 2016. Production of cellulose nanofibrils: A review of recent advances. *Industrial Crops and Products* 93: 2–25.
33. Missoum, K., M.N. Belgacem, and J. Bras. 2013. Nanofibrillated cellulose surface modification: A review. *Materials* 6 (5): 1745–1766.
34. Kumar, R., et al. 2021. A comprehensive review on production, surface modification and characterization of nanocellulose derived from biomass and its commercial applications. *Express Polymer Letters* 15 (2): 104–120.
35. Mokhena, T.C., and M.J. John. 2020. Cellulose nanomaterials: New generation materials for solving global issues. *Cellulose* 27: 1149–1194.
36. Wahlström, R., and A. Suurnäkki. 2015. Enzymatic hydrolysis of lignocellulosic polysaccharides in the presence of ionic liquids. *Green Chemistry* 17 (2): 694–714.
37. Phanthong, P., et al. 2018. Nanocellulose: Extraction and application. *Carbon Resources Conversion* 1 (1): 32–43.
38. Janardhnan, S., and M.M. Sain. 2006. Isolation of cellulose microfibrils–an enzymatic approach. *BioResources* 1 (2): 176–188.
39. Henriksson, M., et al. 2007. An environmentally friendly method for enzyme-assisted preparation of microfibrillated cellulose (MFC) nanofibers. *European Polymer Journal* 43 (8): 3434–3441.
40. López-Rubio, A., et al. 2007. Enhanced film forming and film properties of amylopectin using micro-fibrillated cellulose. *Carbohydrate Polymers* 68 (4): 718–727.
41. Aulin, C., et al. 2010. Self-organized films from cellulose I nanofibrils using the layer-by-layer technique. *Biomacromolecules* 11 (4): 872–882.
42. Xie, H., et al. 2018. Recent strategies in preparation of cellulose nanocrystals and cellulose nanofibrils derived from raw cellulose materials. *International Journal of Polymer Science* 2018.
43. Olszewska, A., et al. 2011. The behaviour of cationic nanofibrillar cellulose in aqueous media. *Cellulose* 18: 1213–1226.
44. Liimatainen, H., et al. 2014. Fabrication of cationic cellulosic nanofibrils through aqueous quaternization pretreatment and their use in colloid aggregation. *Carbohydrate Polymers* 103: 187–192.
45. Ho, T., et al. 2011. Preparation and characterization of cationic nanofibrillated cellulose from etherification and high-shear disintegration processes. *Cellulose* 18: 1391–1406.
46. Fukaya, Y., et al. 2008. Cellulose dissolution with polar ionic liquids under mild conditions: Required factors for anions. *Green Chemistry* 10 (1): 44–46.
47. Vitz, J., et al. 2009. Extended dissolution studies of cellulose in imidazolium based ionic liquids. *Green Chemistry* 11 (3): 417–424.
48. Li, J., et al. 2012. Homogeneous isolation of nanocellulose from sugarcane bagasse by high pressure homogenization. *Carbohydrate Polymers* 90 (4): 1609–1613.
49. Zhu, S., et al. 2006. Dissolution of cellulose with ionic liquids and its application: A mini-review. *Green Chemistry* 8 (4): 325–327.
50. Farooq, A., et al. 2021. Structure and properties of high quality natural cellulose nano fibrils from a novel material Ficus natalensis barkcloth. *Journal of Industrial Textiles* 51 (4): 664–680.
51. Stepanova, G., et al. 1982. The swelling of alkali cellulose in sodium hydroxide solutions. *Fibre Chemistry* 13 (6): 405–407.

52. Belgacem, M.N., and A. Gandini. 2005. The surface modification of cellulose fibres for use as reinforcing elements in composite materials. *Composite Interfaces* 12 (1–2): 41–75.
53. Alemdar, A., and M. Sain. 2008. Isolation and characterization of nanofibers from agricultural residues–Wheat straw and soy hulls. *Bioresource Technology* 99 (6): 1664–1671.
54. Wang, B., and M. Sain. 2007. The effect of chemically coated nanofiber reinforcement on biopolymer based nanocomposites. *BioResources* 2 (3): 371–388.
55. Bhatnagar, A., and M. Sain. 2005. Processing of cellulose nanofiber-reinforced composites. *Journal of Reinforced Plastics and Composites* 24 (12): 1259–1268.
56. Brinchi, L., et al. 2013. Production of nanocrystalline cellulose from lignocellulosic biomass: Technology and applications. *Carbohydrate Polymers* 94 (1): 154–169.
57. Bondeson, D., A. Mathew, and K. Oksman. 2006. Optimization of the isolation of nanocrystals from microcrystalline cellulose by acid hydrolysis. *Cellulose* 13: 171–180.
58. Iskak, N.A.M., N.M. Julkapli, and S.B.A. Hamid. 2017. Understanding the effect of synthesis parameters on the catalytic ionic liquid hydrolysis process of cellulose nanocrystals. *Cellulose* 24: 2469–2481.
59. Man, Z., et al. 2011. Preparation of cellulose nanocrystals using an ionic liquid. *Journal of Polymers and the Environment* 19: 726–731.
60. Cui, S., et al. 2016. Green preparation and characterization of size-controlled nanocrystalline cellulose via ultrasonic-assisted enzymatic hydrolysis. *Industrial Crops and Products* 83: 346–352.
61. Pereira, B., and V. Arantes. 2020. Production of cellulose nanocrystals integrated into a biochemical sugar platform process via enzymatic hydrolysis at high solid loading. *Industrial Crops and Products* 152: 112377.
62. Cao, X., et al. 2012. Cellulose nanowhiskers extracted from TEMPO-oxidized jute fibers. *Carbohydrate Polymers* 90 (2): 1075–1080.
63. Zhang, H., et al. 2020. Extraction and comparison of cellulose nanocrystals from lemon (Citrus limon) seeds using sulfuric acid hydrolysis and oxidation methods. *Carbohydrate Polymers* 238: 116180.
64. Raza, M., and B. Abu-Jdayil. 2022. Cellulose nanocrystals from lignocellulosic feedstock: A review of production technology and surface chemistry modification. *Cellulose* 29 (2): 685–722.
65. Vanderfleet, O.M., and E.D. Cranston. 2021. Production routes to tailor the performance of cellulose nanocrystals. *Nature Reviews Materials* 6 (2): 124–144.
66. Liu, C., et al. 2016. Properties of nanocellulose isolated from corncob residue using sulfuric acid, formic acid, oxidative and mechanical methods. *Carbohydrate Polymers* 151: 716–724.
67. Chen, L., et al. 2016. Highly thermal-stable and functional cellulose nanocrystals and nanofibrils produced using fully recyclable organic acids. *Green Chemistry* 18 (13): 3835–3843.
68. Liu, Y., et al. 2014. A novel approach for the preparation of nanocrystalline cellulose by using phosphotungstic acid. *Carbohydrate Polymers* 110: 415–422.
69. Welton, T. 1999. Room-temperature ionic liquids. Solvents for synthesis and catalysis. *Chemical Reviews* 99 (8): 2071–2084.
70. Earle, M.J., and K.R. Seddon. 2000. Ionic liquids. Green solvents for the future. *Pure and Applied Chemistry* 72 (7): 1391–1398.
71. Jordan, J.H., M.W. Easson, and B.D. Condon. 2020. Cellulose hydrolysis using ionic liquids and inorganic acids under dilute conditions: Morphological comparison of nanocellulose. *RSC Advances* 10 (65): 39413–39424.
72. Ninomiya, K., et al. 2018. Lignocellulose nanofibers prepared by ionic liquid pretreatment and subsequent mechanical nanofibrillation of bagasse powder: Application to esterified bagasse/polypropylene composites. *Carbohydrate Polymers* 182: 8–14.
73. Wang, Y., et al. 2015. Study on nanocellulose by high pressure homogenization in homogeneous isolation. *Fibers and Polymers* 16: 572–578.
74. Zhao, G., et al. 2017. Facile one-pot fabrication of cellulose nanocrystals and enzymatic synthesis of its esterified derivative in mixed ionic liquids. *RSC Advances* 7 (43): 27017–27023.

75. Song, X., et al. 2018. Simultaneous improvement of thermal stability and redispersibility of cellulose nanocrystals by using ionic liquids. *Carbohydrate Polymers* 186: 252–259.
76. Babicka, M., et al. 2020. Preparation of nanocellulose using ionic liquids: 1-propyl-3-methylimidazolium chloride and 1-ethyl-3-methylimidazolium chloride. *Molecules* 25 (7): 1544.
77. Sharma, V., P. Nargotra, and B.K. Bajaj. 2019. Ultrasound and surfactant assisted ionic liquid pretreatment of sugarcane bagasse for enhancing saccharification using enzymes from an ionic liquid tolerant Aspergillus assiutensis VS34. *Bioresource Technology* 285: 121319.

Chapter 3
Characteristics of Deep Eutectic Solvents

Abstract Deep eutectic solvents (DESs) exhibit unique physical properties that align with the criteria for green solvents, including safety, non-toxicity (particularly for natural DESs), non-flammability, non-volatility, thermal stability, sustainability, biodegradability, and cost-effectiveness. DES has gained significant attention in the field of green chemistry over the past seventeen years, resulting in numerous publications exploring various research areas and diverse applications. These solvents offer promising alternatives to traditional organic solvents, primarily due to their simple preparation using readily available and natural compounds. Additionally, DESs demonstrate a high degree of tunability, allowing for customization to specific requirements. In this chapter, we provide a comprehensive overview of the properties of deep eutectic solvents including melting point, biodegradability, density, and surface tension. These properties can be controlled by factors such as the choice of constituent compounds, their molar ratio, temperature, and water content.

3.1 Introduction

Deep eutectic solvents represent a novel class of environmentally friendly solvents composed of two or more components. These solvents exhibit liquid phase behavior at ambient temperature through hydrogen bond interactions, resulting in a eutectic composition with a lower melting point than the individual components. DESs offer several advantages over traditional solvents, including renewability, biodegradability, and non-toxicity, making them a compelling alternative for various chemical applications [1]. The unique properties of DESs have sparked considerable research interest in their potential use as solvents in diverse fields such as electrochemistry, organic synthesis, and catalysis. Additionally, DESs are well-suited for sustainable chemistry, a critical objective in minimizing the environmental impact of chemical processes, owing to their easy biodegradability and reduced toxicity [2]. In recent years, there has been a growing interest among researchers in deep eutectic solvents as potential green solvents, aiming to replace the harmful and toxic volatile organic solvents that have been widely used for centuries. Abbott et al. were the pioneers in reporting the

A. Farooq et al., *Deep Eutectic Solvents in the Textile Industry*,
https://doi.org/10.1007/978-981-97-6433-4_3

synthesis of DESs, utilizing choline chloride as a hydrogen bond acceptor (HBA) and urea as a hydrogen bond donor (HBD). These DESs exhibited remarkable physical properties, including a high depression melting point, and represented a molten liquid salt at room temperature due to strong interactions between the HBA and HBD components [3]. Deep eutectic solvents have demonstrated excellent potential in the dissolution and pretreatment of cellulose, distinguishing them from ionic liquids (ILs) despite their similar chemical and physical properties. DESs are typically derived from inorganic salts, but when natural components such as sugars, amino acids, or organic acids are utilized, they are referred to as natural deep eutectic solvents (NADES). NADES, being a naturally occurring metabolite, offers inherent advantages over traditional solvents. It exhibits excellent eco-friendliness, environmental compatibility, non-toxicity, cost-effectiveness, ease of utilization and synthesis, recyclability, and biocompatibility. In this chapter, we delve into the characteristics of deep eutectic solvents with a specific focus on their biodegradability, density, and melting point. Biodegradability is a crucial aspect of DESs, as it determines their environmental impact and sustainability [4]. Density is another important property of DESs, which affects their phase behavior and applications in various industries [5, 6]. We explore the factors that influence the density of DESs, such as the composition and molecular interactions between the components. Additionally, we discuss the melting point of DESs, which plays a significant role in their practical usage. The melting point is influenced by the eutectic composition and the strength of the hydrogen bonding between the components [7, 8]. Understanding the biodegradability, density, and melting point of DESs provides valuable insights into their potential applications in fields such as green chemistry, materials science, and biotechnology. Figure 3.1 shows the various characteristics of deep eutectic solvents.

3.2 Properties of Deep Eutectic Solvents

3.2.1 Melting Point

Research on DESs has mainly focused on defining the melting point (phase transition point) of eutectic systems, which has been known for many years as a crucial quality in the preparation of deep eutectic solvents. Generally, every report or study on DES also includes the phase transition point which is named as the melting point of a deep eutectic solvent system. Melting point characterization is generally done by using differential scanning calorimetry (DSC), accompanied by an optical microscope and other relevant instruments. Despite this, reports found that deep eutectic solvents usually do not have a real melting point, but rather display the glass transition temperature at which they are most stable [9, 10]. Researchers are trying to find the exact reason for this phenomenon; however, studies considered the hydrogen bonding to hinder the deep eutectic solvent crystallization. The strong and excellent crosslinking between hydrogen bond acceptor and hydrogen bond donor is causing

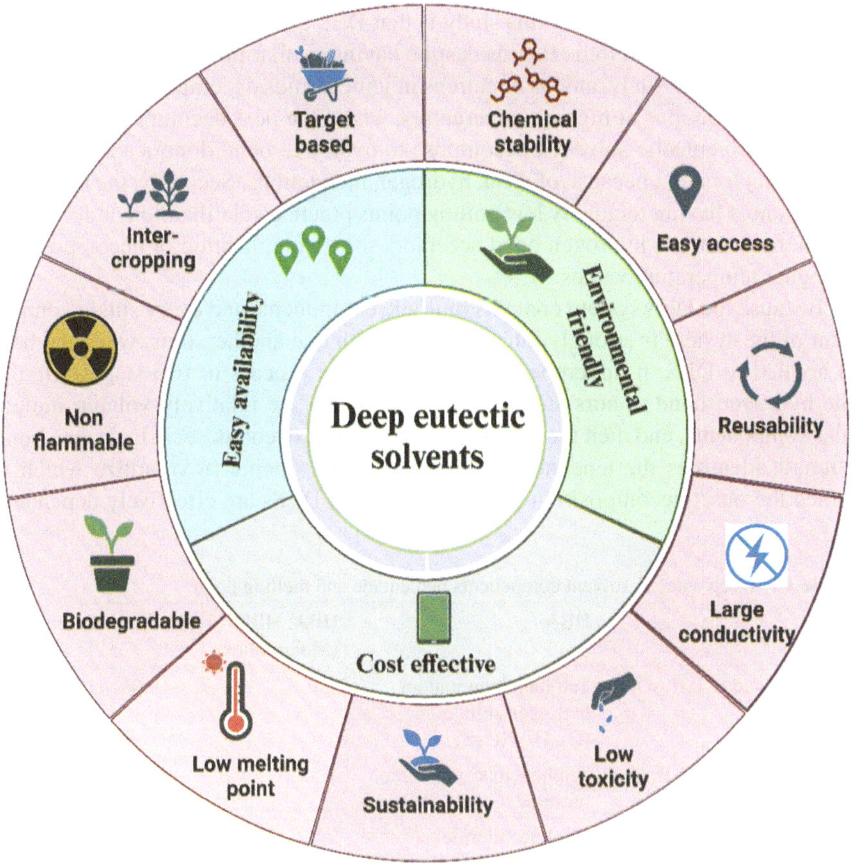

Fig. 3.1 Characteristics of deep eutectic solvents

the low melting point of eutectics. Deep eutectic solvents having melting points lower than 50 °C are widely explored by scientists because of their safe, low-cost, and easy-to-use characteristics in industrial field applications, and they are named low-transition temperature mixtures [11]. However, organic salt's molar ratio, alkyl chain length, and hydrogen bond donor-imposed a great impact on the melting point of deep eutectic solvents as depicted in Table 3.1. As DESs are most similar to ionic liquids, thus they are considered as fewer volatile solvents. Nevertheless, researchers are curious to carefully study their thermal stability properties, that's why their research revealed that deep eutectic solvents have low however detectable vapor pressure under pervasive conditions [12]. The core of vapor pressure has been considered to be the reason for the volatilization of a single or every component in DES. Thus, in a study, a systemic analysis of the thermal stability of about 40 DESs was deeply and carefully evaluated by utilizing thermal gravimetric analysis, and as a result, the onset decomposition temperature was attained [13]. One of the most important and comprehensive findings from this study is that DES exhibits a completely different thermal behavior from ionic liquids despite having similar properties. According to the findings of the study, anions or cations in ionic liquids decompose first, followed by the opposite ions at higher temperatures, where the next decomposition occurs. But for deep eutectic solvents, decompose to hydrogen bond donors and hydrogen bond acceptors first because of weak hydrogen interactions. Secondly, the hydrogen bond donors having relatively low boiling points practice volatilization or decomposition, however, the hydrogen bond acceptors show volatilization or decomposition at higher temperature values.

Because the DES system contains multiple components and every single component of the system is strongly interconnected with one another, thus, when the heat is applied to DES, it undergoes the decomposition process in two stages. Firstly, the hydrogen bond donors start to evaporate which are relatively volatile molecular components, and then the remaining components decompose. A hydrogen bond strength identifies the tendency for hydrogen bond donors to volatilize which is called the onset decomposition temperature. Thus, DESs are effectively dependent

Table 3.1 Deep eutectic solvent components percentage and melting point

HBD	HBA	HBA: HBD (Molar ratios)	Tm/°C	References
Benzilic acid ($C_{14}H_{12}O_3$)	Tetrabutylammonium hydrogen sulfate ($C_{16}H_{37}NO_4S$)	1:1	−19.06	[14]
Ethylene glycol ($C_2H_6O_2$)	Triphenylphosphunium bromide ($C_{18}H_{16}BrP$)	1:5	−48	[15]
Glycerol ($C_3H_8O_3$)	Choline chloride [$(CH_3)3NCH_2CH_2OH$]Cl	1:2	−1	[16]
Ethylene glycol ($C_2H_6O_2$)	Triphenylphosphunium bromide ($C_{18}H_{16}BrP$)	1:3	−46	[15]
Glycerol ($C_3H_8O_3$)	Choline chloride [$(CH_3)3NCH_2CH_2OH$]Cl	1:3	1.7	[16]
Benzilic acid ($C_{14}H_{12}O_3$)	Tetraethylammonium bromide ($C_8H_{20}NBr$)	1:1	−11.06	[14]
Glycerol ($C_3H_8O_3$)	Choline chloride [$(CH_3)3NCH_2CH_2OH$]Cl	1:4	2	[16]
Pyrogallol ($C_6H_6O_3$)	Choline chloride [$(CH_3)3NCH_2CH_2OH$]Cl	1:2	−50.9	[16]
Benzilic acid ($C_{14}H_{12}O_3$)	Tetraethylammonium bromide ($C_8H_{20}NBr$)	1:1	−7.54	[14]
Ethylene glycol ($C_2H_6O_2$)	Triphenylphosphunium bromide ($C_{18}H_{16}BrP$)	1:4	−50	[15]
Pyrogallol ($C_6H_6O_3$)	Choline chloride [$(CH_3)3NCH_2CH_2OH$]Cl	1:1	−61.9	[16]

on the thermal stability of hydrogen bond donors to maintain their thermal stability. Additionally, in other studies, thermogravimetric analysis (TGA), Fourier transition electron microscope (FTIR), and 2D-IR have been used to evaluate the volatility and long-term volatility of deep eutectic solvents [17, 18]. The findings of the studies revealed that the DESs show volatile properties under ambient temperature and pressure, which is why it is very important at the industrial scale to carefully check the tightness of the container.

Deep eutectic solvents containing two components; most notably choline chloride and metal chlorides [19] or hydrogen bond donors including various types of amides, acids, and alcohols [20, 21] have excellent low melting temperature and these DES created by various forming complex creates homogenous liquid of ions with significant decreased freezing points [21]. As compared to traditionally used imidazolium-type ionic liquids, deep eutectic solvents especially choline chloride based DES easily be prepared even at low cost with excellent purity including low toxic issues [20]. DESs show outstanding low melting point solvent properties and has been reported to dissolve various compounds including synthetic polymers, metals, and carbohydrates [22–24]. Thus, DES can be an excellent choice for textile materials coating and etching by applying functional materials to textile fibers and fabrics in deep eutectic-based solvents medium to obtain tremendous functional properties according to the requirements such as antibacterial, ultraviolet, hydrophobic, hydrophilic, and so on. Thus by using deep eutectic solvents a proposed structure can be drawn in case of using choline chloride based DES for functional modification of textile materials like cotton fabric through 3,3',4,4'-Benzophenonetetracarboxylic dianhydride as shown in Fig. 3.2, which reveals that reaction of 3,3',4,4'-Benzophenonetetracarboxylic dianhydride with cellulose molecule can be done through a classical nucleophilic acyl substitution reaction between anhydrides and hydroxyl groups in anhydroglucose unit to form ester (part 1), and part 2 shows possible reaction mechanism between 3,3',4,4'-Benzophenonetetracarboxylic dianhydride and urea as illustrated in Fig. 3.2.

3.2.2 Biodegradability of DES

In recent years, DES has been extensively used in the production of eco-friendly solvents for a variety of applications due to the development of a new generation of these solvents very recently. However, the ecological impact of this industry has not been thoroughly studied yet. Thus, deep eutectic solvent's biodegradability was evaluated in a report for the first time by using DES mixtures prepared by choline chloride with the glycerol, glucose, and oxalic acid in 1:1, 1:2, and 2:1 molar ratio, through an aerobic biodegradability test [25]. Radosevic et al. revealed that after a standard period test (28 days), the biodegradability of all the DES samples exceeded 60%, with the highest biodegradation levels being reported for glycerol, glucose, and oxalic acid at 96%, 84%, and 68%, respectively [25]. The excellent neutrality of each hydrogen bond donor and acceptor is generally responsible for

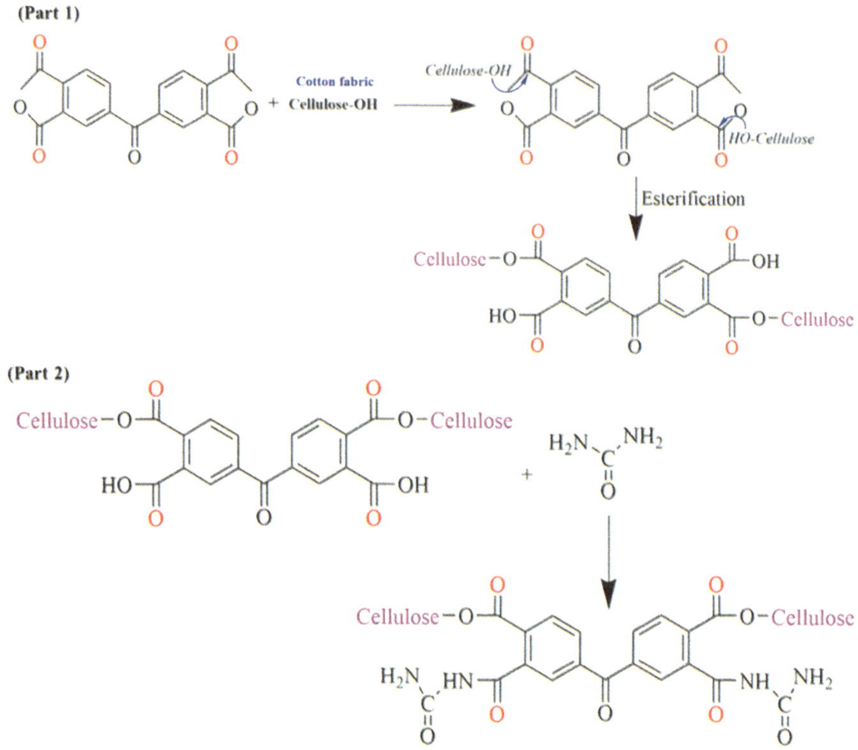

Fig. 3.2 Proposed reaction mechanism during chemical modification of textile cotton fabric through esterification

DES's biodegradability, as a result, a naturally occurring DES mixture is created, which can be easily metabolized by bacteria as part of the reactions. Therefore, it is possible to easily measure the biodegradability of DES by using both the closed bottle test method as well as respirometric techniques. Meanwhile, a study was conducted to examine the biodegradability of DES mixtures prepared by various constituents with a molar ratio of 1:1 using choline chloride, ethylene glycol, acetamide, and glycerol. The results revealed that the DESs mixture containing urea and acetamide-based attained about 80% biodegradability level, however, other components making DESs showed lower biodegradability levels (less than 60%) [26]. DESs are generally considered to be "green" because of their eco-friendly chemical composition (consisting of ammonium salts and hydrogen-bond donors) [27]. A biodegradability test was conducted using a closed bottle test method against Escherichia coli (a bacterium), Allium sativum (garlic, a plant), and the hydra (an invertebrate) to calculate the level of biodegradability [26]. The study involves the evaluation of glycerol biodegradability using various experimental factors including different molar ratios and reaction conditions. Several reports have been devoted to the green properties of

DESs. Some of the studies were conducted based on the pharmaceutically acceptable profile which has been mentioned in the material safety data sheets of individual constituents [28]. In a study, an evaluation of the DES component's impact on performance of solvent on fish and human cell line was done by using three various types of ChCl-based DESs [29]. Cvjetko Bubalo et al. [30] examined the biodegradability of DES and mentioned that the percentage of yeast cells within aqueous solution of ChCl: EG-DES were identical to those within a conventional buffer. Similarly, Radošević et al. [29] mentioned in a study that all deep eutectic solvents tested (ChC: Gly, ChCl: Glc, ChCl: OA) can be categorized as biodegradable having biodegradability level 68–96%. However, in a report, Zhao et al. [31] showed the biodegradability of DESs (different molar ratios of 19 choline chloride-based DES) by using standard techniques and mentioned the biodegradability levels as 69.3%, 95.9%, and 97.1% for triethylene glycol, glycerol, and urea, respectively, Table 3.2 shows the biodegradability level against various types of organisms.

Table 3.2 Amine, sugar, alcohol and acid based DESs biodegradability

Hydrogen bond acceptors (HBA)	Hydrogen bond donors (HBD)	Molar ratios	Biodegradability level percentage (%)	Test organisms	Ref
Ethylene glycol	Choline chloride	2:1	75	Bacteria (A. fisheri), crustaceans (D. magna), and algae (S. capricornatum)	[32]
Urea	Choline chloride	2:1	63		
Glycerol	Choline chloride	2:1	75		
Malonic acid	Ethylene ammonium chloride	1:1	61	Aspergillus niger, Cyprinus carpio fish	[33]
Malonic acid	Choline chloride	1:1	76.3	Staphylococcus aureus and Listeria monocytogenes, Escherichia coli and Salmonella enteritidis	[31]
Ethylene glycol	Choline chloride	2:1	81.9		
Fructose/water	Choline chloride	5:2:5	93.7		
Triethylene glycol	Choline chloride	2:1	69.3		
Glycerol	Choline chloride	2:1	95.9		
Urea	Choline chloride	2:1	97.1		
Glycerol	Choline chloride	2:1	100	Wastewater microorganisms	[25]

3.2.3 Density of Deep Eutectic Solvents

In the DES system, density is one of the most critical factors that can provide information regarding intermolecular interaction as well as crosslinking between molecules within the material. Normally, DESs have been exhibiting much higher densities as compared to water for example at 20 °C ethaline and glyceline densities were reported to show higher density values [34]. According to the Basaiahgari et al. study, DESs based on ethylene glycol display lower density values as compared to glycerol-based DES during the density measurements of ethylene, diethylene, triethylene glycol, and glycerol as HBDs and benzylammonium chloride salts as HBAs [35]. The increase in density is a result of increasing the number of –OH functional groups on the hydrogen bond donor resulting in the formation of more hydrogen bonds, which will probably reduce the available free volume. DESs have varying densities depending on how the molecules are arranged and packed within them, which is highly affected by the presence of holes and liquids in deep eutectic solvents. That's why several studies reveal the higher density of the choline chloride/urea system as compared to acetylcholine chloride/urea because of the presence of free space and holes in acetylcholine chloride as acknowledged in [16, 36, 37]. In many cases, however, it is the chemical composition and molecular ratio of HBA and HBD that is the most important factor in increasing the density of the DES system. Accordingly, Shafie et al. [38] examined the molar ratio of choline chloride to citric acid-based DESs to investigate the possible mechanism. Their results revealed that with the increase of the amount of choline chloride in the constant value of citric acid, a reduction in density was noted. While, with the increase of citric acid content relative to choline chloride, the density values increase as well. Deep eutectic solvent densities alter with the temperature variation in which density values are reported to decrease with the increase of temperature. For example, the DES system shows a variation in density dramatically, the reduction in density of reline was inspected with the increase of temperature is a result of an increase of ionic motion and free volume in the liquid DES [39]. DES density variations as a function of temperature were also predicted by Shahbaz et al. [40]. Similarly, when the alkyl chain length of the cation increases, the density of DES is greatly reduced because of an obvious increase in chain length [41, 42]. According to our knowledge, there have been few research publications exclusively on DES's density. However, some experiments have been done on density in front of application experiments [43, 44]. As density is a physical property of DES thus it cannot be considered meaningless rather these properties are more significant in industrial research. Table 3.3 shows the various formations of DES constituents and densities of the prepared materials.

Table 3.3 DES formation and densities at a constant temperature

HBA	HBD	Temperature (°C)	DES formation (Molar ratio)	Density (g/cm^3)	References
Acetylcholine chloride	Levulinic acid	25	1:2	1.14	[45]
Acetylcholine chloride	Urea	25	1:2	1.20	[46]
Tetraethyl ammonium bromide	Pyrogallol	25	1:2	1.30	[16]
Tetraproyl ammonium bromide	Pyrogallol	25	1:1	1.20	
Tetraproyl ammonium bromide	Pyrogallol	25	1:1	1.13	
Octylammonium chloride	Decanoinc acid	25	1:2	0.89	[47]
N,N-diethylenethanol ammonium chloride	Ethylene glycol	25	1:4	1.1	[48]
Choline chloride	Phenol	25	1:2	1.09	[49]
Choline chloride	Phenol	25	1:3	1.09	[21]
Choline chloride	Levulinic acid	25	1:3	1.13	[50]
Choline chloride	Levulinic acid	25	1:4	1.13	[51]
Choline chloride	Glucose	25	2:1	1.24	[52]
Choline chloride	Glucose	25	2:3	1.26	[53]
Choline chloride	Fructose	25	1:1	1.27	[53]
Choline chloride	Fructose	25	2:1	1.27	[52]

3.2.4 *Surface Tension of DES*

Generally, surface tension (ST) of any material has been reported as the exact measure of energy that enhances the surface area of a material, thus, ST of DES is a physical property that shows that changes in composition and temperature dramatically affect the intensity of interactions between DES constituents (HBD and HBA) [54, 55]. However, due to intermolecular interactions between molecules, ST effects are more prevalent in liquids [56], in which the analysis is generally done by using Wilhelmy plates, a Du Noüy ring, and pendant drops [57–59]. DES surface tension depends on component types and their nature as well as reaction conditions as shown in Table 3.4.

Table 3.4 Surface tension of DES at constant room temperature

Hydrogen bond acceptor (HBA)	Hydrogen bond donor (HBD)	Relative molar ratio	Surface tension ($mN.m^{-1}$)	References
Betaine	Lactic acid	1:2	42.4	[60]
Betaine	Lactic acid	1:2	42.6	[60]
Choline chloride	Ethylene glycol	1:2	48.91	[61]
Choline chloride	Lactic acid	1:4	44.4	[60]
Choline chloride	Lactic acid	1:2	47.4	[60]
Choline chloride	Phenylacetic acid	1:2	57.96	[46]
Polyethylene glycol	Lactic acid	2:1	41.3	[60]
Choline chloride	Glycerol	1:2	55.4	[62]
Choline chloride	Ethylene glycol	1:2	48.91	[63]
Polyethylene glycol	Benzamide	2:1	42.6	[60]
Polyethylene glycol	Citric acid	4:1	42.6	[60]
Butylammonium bromide	Glycerol	1:2	44.9	[64]
Polyethylene glycol	Thiourea	2:1	43.8	[60]
Ethylammonium bromide	Glycerol	1:2	57.6	[64]
Polyethylene glycol	Oxalic acid	2:1	44.1	[60]
Propylammonium bromide	Glycerol	1:2	51.7	[64]
Choline chloride	Urea	1:2	52	[63]
Polyethylene glycol	Glutaric acid	2:1	43.3	[60]

3.3 Conclusion

In conclusion, deep eutectic solvents offer excellent advantages that make them desirable for various applications. Their biodegradability ensures minimal environmental impact and aligns with the principles of green chemistry. Furthermore, DESs exhibit low melting points, which enhance their usability and efficiency in different processes. The ability of DESs to vary in density according to reaction conditions, such as time and temperature, provides flexibility and control in chemical reactions. This characteristic allows for tailored solvent properties to match specific reaction requirements. Overall, the combination of biodegradability, low melting points, and density variation in DESs makes them a promising choice for sustainable and efficient solvent applications.

References

1. Zhang, Q., et al. 2012. Deep eutectic solvents: Syntheses, properties and applications. *Chemical Society Reviews* 41 (21): 7108–7146.
2. Coutinho, J.A., and S.P. Pinho. 2017. Special issue on deep eutectic solvents: a foreword. *Fluid Phase Equilibria* 448 (1).
3. Ni, Y., et al. 2019. Deep eutectic solvent (DES) as both solvent and catalyst for oxidation of furfural to maleic acid and fumaric acid. *Green Chemistry* 21 (5): 1075–1079.
4. Azizi, N., and E. Batebi. 2012. Highly efficient deep eutectic solvent catalyzed ring opening of epoxides. *Catalysis Science & Technology* 2 (12): 2445–2448.
5. Gu, T., et al. 2014. Deep eutectic solvents as novel extraction media for phenolic compounds from model oil. *Chemical Communications* 50 (79): 11749–11752.
6. Abbott, A.P., et al. 2014. Evaluating water miscible deep eutectic solvents (DESs) and ionic liquids as potential lubricants. *Green Chemistry* 16 (9): 4156–4161.
7. Omar, K.A., and R. Sadeghi. 2020. Novel diglycolic acid-based deep eutectic solvents and their applications as a rust remover. *Journal of Molecular Liquids* 312: 113380.
8. Xiong, Z., et al. 2019. Ultrasound-assisted deep eutectic solvent as green and efficient media for the extraction of flavonoids from Radix scutellariae. *New Journal of Chemistry* 43 (2): 644–650.
9. Li, N., et al. 2016. Development of green betaine-based deep eutectic solvent aqueous two-phase system for the extraction of protein. *Talanta* 152: 23–32.
10. Florindo, C., L. Branco, and I. Marrucho. 2017. Development of hydrophobic deep eutectic solvents for extraction of pesticides from aqueous environments. *Fluid Phase Equilibria* 448: 135–142.
11. Rodriguez Rodriguez, N., L. Machiels, and K. Binnemans. 2019. p-Toluenesulfonic acid-based deep-eutectic solvents for solubilizing metal oxides. *ACS Sustainable Chemistry & Engineering* 7 (4): 3940–3948.
12. Sharma, M., et al. 2013. Dissolution of α-chitin in deep eutectic solvents. *RSC Advances* 3 (39): 18149–18155.
13. Rastegarifard, F., K. Ghanemi, and M. Fallah-Mehrjardi. 2017. A deep eutectic solvent-based extraction method for fast determination of Hg in marine fish samples by cold vapor atomic absorption spectrometry. *Analytical Methods* 9 (39): 5741–5748.
14. Omar, K.A., and R. Sadeghi. 2020. Novel benzilic acid-based deep-eutectic-solvents: Preparation and physicochemical properties determination. *Fluid Phase Equilibria* 522: 112752.
15. Shahbaz, K., et al. 2011. Using deep eutectic solvents based on methyl triphenyl phosphunium bromide for the removal of glycerol from palm-oil-based biodiesel. *Energy & Fuels* 25 (6): 2671–2678.
16. Omar, K.A., and R. Sadeghi. 2021. Novel deep eutectic solvents based on pyrogallol: Synthesis and characterizations. *Journal of Chemical & Engineering Data* 66 (5): 2088–2095.
17. Chen, Y., et al. 2020. Vaporization enthalpy, long-term evaporation and evaporation mechanism of polyethylene glycol-based deep eutectic solvents. *New Journal of Chemistry* 44 (22): 9493–9501.
18. Chen, Y., et al. 2019. Volatility of deep eutectic solvent choline chloride: N-methylacetamide at ambient temperature and pressure. *Industrial & Engineering Chemistry Research* 58 (17): 7308–7317.
19. Disale, S.T., et al. 2012. Choline chloride· 2ZnCl2 ionic liquid: An efficient and reusable catalyst for the solvent free Kabachnik-Fields reaction. *Tetrahedron Letters* 53 (18): 2277–2279.
20. Abbott, A.P., et al. 2003. Novel solvent properties of choline chloride/urea mixtures. *Chemical Communications* 1: 70–71.
21. Abbott, A.P., et al. 2004. Deep eutectic solvents formed between choline chloride and carboxylic acids: Versatile alternatives to ionic liquids. *Journal of the American Chemical Society* 126 (29): 9142–9147.

22. Abbott, A.P., and K.J. McKenzie. 2006. Application of ionic liquids to the electrodeposition of metals. *Physical Chemistry Chemical Physics* 8 (37): 4265–4279.
23. Gu, C., X. Xu, and J. Tu. 2010. Fabrication and wettability of nanoporous silver film on copper from choline chloride-based deep eutectic solvents. *The Journal of Physical Chemistry C* 114 (32): 13614–13619.
24. Kwon, M.-Y., H.-M. Choi, and C.-J. Hong. 2009. Absorbency characterization of polyester fabrics modified with nitrogen compounds. *Textile Science and Engineering* 46 (3): 129–135.
25. Radošević, K., et al. 2015. Evaluation of toxicity and biodegradability of choline chloride based deep eutectic solvents. *Ecotoxicology and Environmental Safety* 112: 46–53.
26. Wen, Q., et al. 2015. Assessing the toxicity and biodegradability of deep eutectic solvents. *Chemosphere* 132: 63–69.
27. Khandelwal, S., Y.K. Tailor, and M. Kumar. 2016. Deep eutectic solvents (DESs) as eco-friendly and sustainable solvent/catalyst systems in organic transformations. *Journal of Molecular Liquids* 215: 345–386.
28. Singh, B.S., H.R. Lobo, and G.S. Shankarling. 2012. Choline chloride based eutectic solvents: Magical catalytic system for carbon–carbon bond formation in the rapid synthesis of β-hydroxy functionalized derivatives. *Catalysis Communications* 24: 70–74.
29. Radošević, K., M.C. Bubalo, V.G. Gaurina Srček, D. Grgas, T.L. Landeka Dragičević, I.R. Radojčić Redovniković. 2015. Evaluation of toxicity and biodegradability of choline chloride based deep eutectic solvents. *Ecotoxicology and Environmental Safety* 112: 46–53.
30. Bubalo, M.C., et al. 2015. Baker's yeast-mediated asymmetric reduction of ethyl 3-oxobutanoate in deep eutectic solvents. *Process Biochemistry* 50 (11): 1788–1792.
31. Zhao, B.-Y., et al. 2015. Biocompatible deep eutectic solvents based on choline chloride: Characterization and application to the extraction of rutin from Sophora japonica. *ACS Sustainable Chemistry & Engineering* 3 (11): 2746–2755.
32. Lapeña, D., et al. 2021. Ecotoxicity and biodegradability of pure and aqueous mixtures of deep eutectic solvents: Glyceline, ethaline, and reline. *Environmental Science and Pollution Research* 28 (7): 8812–8821.
33. Juneidi, I., M. Hayyan, and M.A. Hashim. 2015. Evaluation of toxicity and biodegradability for cholinium-based deep eutectic solvents. *RSC Advances* 5 (102): 83636–83647.
34. Mjalli, F.S., and H. Mousa. 2017. Viscosity of aqueous ionic liquids analogues as a function of water content and temperature. *Chinese Journal of Chemical Engineering* 25 (12): 1877–1883.
35. Basaiahgari, A., S. Panda, and R.L. Gardas. 2018. Effect of ethylene, diethylene, and triethylene glycols and glycerol on the physicochemical properties and phase behavior of benzyltrimethyl and benzyltributylammonium chloride based deep eutectic solvents at 283.15–343.15 K. *Journal of Chemical & Engineering Data* 63 (7): 2613–2627.
36. Omar, K.A., and R. Sadeghi. 2021. Novel Nonanol-Based deep eutectic solvents: Thermophysical properties and their applications in Liquid-Liquid extraction and amino acid detection. *Journal of Molecular Liquids* 336: 116359.
37. Abbott, A.P., et al. 2007. Eutectic-based ionic liquids with metal-containing anions and cations. *Chemistry–A European Journal* 13 (22): 6495–6501.
38. Shafie, M.H., R. Yusof, and C.-Y. Gan. 2019. Synthesis of citric acid monohydrate-choline chloride based deep eutectic solvents (DES) and characterization of their physicochemical properties. *Journal of Molecular Liquids* 288: 111081.
39. El Achkar, T., H. Greige-Gerges, and S. Fourmentin. 2021. Basics and properties of deep eutectic solvents: A review. *Environmental Chemistry Letters* 19 (4): 3397–3408.
40. Shahbaz, K., et al. 2011. Prediction of deep eutectic solvents densities at different temperatures. *Thermochimica Acta* 515 (1–2): 67–72.
41. García, G., et al. 2015. Deep eutectic solvents: Physicochemical properties and gas separation applications. *Energy & Fuels* 29 (4): 2616–2644.
42. Montalbán, M., et al. 2015. Effect of temperature, anion, and alkyl chain length on the density and refractive index of 1-alkyl-3-methylimidazolium-based ionic liquids. *Journal of Chemical & Engineering Data* 60 (7): 1986–1996.

43. Santos, F., M.I. PS Leitão, and A.R. C. Duarte. 2018. Properties of therapeutic deep eutectic solvents of L-arginine and ethambutol for tuberculosis treatment. *Molecules* 24 (1): 55.
44. Mano, F., et al. 2017. Production of electrospun fast-dissolving drug delivery systems with therapeutic eutectic systems encapsulated in gelatin. *An Official Journal of the American Association of Pharmaceutical Scientists* 18 (7): 2579–2585.
45. Deng, D., et al. 2016. Investigation of solubilities of carbon dioxide in five levulinic acid-based deep eutectic solvents and their thermodynamic properties. *The Journal of Chemical Thermodynamics* 103: 212–217.
46. Abbott, A.P., G. Capper, and S. Gray. 2006. Design of improved deep eutectic solvents using hole theory. *Chemphyschem: A European Journal Of Chemical Physics And Physical Chemistry* 7 (4): 803–806.
47. van Osch, D.J., et al. 2015. Hydrophobic deep eutectic solvents as water-immiscible extractants. *Green Chemistry* 17 (9): 4518–4521.
48. Shahbaz, K., et al. 2012. Densities of ammonium and phosphonium based deep eutectic solvents: Prediction using artificial intelligence and group contribution techniques. *Thermochimica Acta* 527: 59–66.
49. Zhu, J., et al. 2017. Physicochemical properties of deep eutectic solvents formed by choline chloride and phenolic compounds at T = (293.15 to 333.15) K: The influence of electronic effect of substitution group. *Journal of Molecular Liquids* 232: 182–187.
50. Lu, M., et al. 2015. Solubilities of carbon dioxide in the eutectic mixture of levulinic acid (or furfuryl alcohol) and choline chloride. *The Journal of Chemical Thermodynamics* 88: 72–77.
51. Wang, Y., et al. 2016. Roles of a hydrogen bond donor and a hydrogen bond acceptor in the extraction of toluene from n-heptane using deep eutectic solvents. *Green Chemistry* 18 (10): 3089–3097.
52. Mjalli, F.S., K. Shahbaz, and I.M. AlNashef. 2015. Modified Rackett equation for modelling the molar volume of deep eutectic solvents. *Thermochimica Acta* 614: 185–190.
53. Wahab, A., and S. Mahiuddin. 2004. Isentropic compressibility, electrical conductivity, shear relaxation time, surface tension, and Raman spectra of aqueous zinc nitrate solutions. *Journal of Chemical & Engineering Data* 49 (1): 126–132.
54. Grudniewska, A., E. de Melo, and A. Chan, Gniłka, R. ła., Boratynski, F., Matharu, AS. 2018. Enhanced protein extraction from oilseed cakes using glycerol-choline chloride deep eutectic solvents: A biorefinery approach. *ACS Sustainable Chemistry & Engineering* 6 (11): 15791–15800.
55. Bai, C., Q. Wei, and X. Ren. 2017. Selective extraction of collagen peptides with high purity from cod skins by deep eutectic solvents. *ACS Sustainable Chemistry & Engineering* 5 (8): 7220–7227.
56. Srivastava, S. 2020. Knoevenagel condensation and Michael addition in bio-renewable deep eutectic solvent: Facile synthesis of a library of Bis-enol derivatives. *ChemistrySelect* 5 (2): 799–803.
57. Vidal, C., et al. 2016. Exploiting deep eutectic solvents and organolithium reagent partnerships: Chemoselective ultrafast addition to imines and quinolines under aerobic ambient temperature conditions. *Angewandte Chemie International Edition* 55 (52): 16145–16148.
58. Mulks, F.F., et al. 2020. Ambient Moisture Accelerates Hydroamination Reactions of Vinylarenes with Alkali-Metal Amides under Air. *Angewandte Chemie International Edition* 59 (43): 19021–19026.
59. Punzi, A., et al. 2017. Pd-catalyzed thiophene–aryl coupling reaction via C-H bond activation in deep eutectic solvents. *Organic Letters* 19 (18): 4754–4757.
60. Chen, Y., et al. 2019. Surface tension of 50 deep eutectic solvents: Effect of hydrogen-bonding donors, hydrogen-bonding acceptors, other solvents, and temperature. *Industrial & Engineering Chemistry Research* 58 (28): 12741–12750.
61. Marcus, Y. 2018. Estimation of the critical temperatures of some more deep eutectic solvents from their surface tensions. *Advances in Materials Science and Engineering* 2018.

62. Tariq, M., et al. 2012. Surface tension of ionic liquids and ionic liquid solutions. *Chemical Society Reviews* 41 (2): 829–868.
63. D'Agostino, C., et al. 2011. Molecular motion and ion diffusion in choline chloride based deep eutectic solvents studied by 1 H pulsed field gradient NMR spectroscopy. *Physical Chemistry Chemical Physics* 13 (48): 21383–21391.
64. Chen, Z., et al. 2017. Effect of cation alkyl chain length on surface forces and physical properties in deep eutectic solvents. *Journal of Colloid and Interface Science* 494: 373–379.

Chapter 4
Formation and Potential Applications of Deep Eutectic Solvents

Abstract This chapter aims to provide a selection of examples showcasing the diverse range of applications proposed for deep eutectic solvents within the relatively short timeframe since their first research publication in 2003. Since its introduction, DES has attracted significant attention and has been extensively studied across various fields. Their unique solvent properties, enabling the dissolution of a wide range of solutes, coupled with their environmentally friendly nature, have positioned them as promising alternatives to conventional solvents. This chapter emphasis the various applications of deep eutectic solvents that were proposed over the years after the discovery of deep eutectic solvents.

4.1 Introduction

Deep eutectic solvents are widely acknowledged as a novel category of solvents that bear resemblance to ionic liquids (ILs) in terms of their characteristics and properties. However, it is important to note that DESs and ILs are distinct solvent types. DESs are composed of eutectic mixtures of Lewis or Brønsted acids and bases, which can encompass various anionic and/or cationic species. On the other hand, ILs are predominantly composed of discrete anionic and cationic species. Although the terms DES and IL have been used interchangeably in some literature, they represent different solvent systems [1].

While DESs share physical properties with traditional ILs, such as low melting points and high thermal stability, their chemical properties suggest distinctive application areas. DESs' unique composition of acids and bases in eutectic mixtures leads to different chemical interactions and reactivity. This distinction implies that DESs may find applications that significantly differ from those of ILs. The presence of diverse anionic and cationic species in DESs offers opportunities for tailoring solvent properties and specific applications across various fields [2–4].

A. Farooq et al., *Deep Eutectic Solvents in the Textile Industry*,
https://doi.org/10.1007/978-981-97-6433-4_4

To address the limitations of ionic liquids (ILs) while retaining their outstanding properties, a new generation of solvents called deep eutectic solvents is being used alternatively [5]. DESs are often referred to as designer solvents, as they offer the ability to tailor their properties for specific applications. These solvents are typically prepared by combining a hydrogen bond acceptor (HBA), such as a quaternary ammonium salt, with a hydrogen bond donor (HBD), such as a carboxylic acid. The interaction between the HBA and HBD, which involves charge delocalization through hydrogen bonding, leads to a depression in the melting point of the resulting DES [1]. DESs possess several advanced characteristics that make them promising for scientific innovation. They can be easily prepared using simple methods, making them cost-effective compared to ILs. Additionally, DESs are biodegradable, further enhancing their environmental friendliness. The choice of HBA is often quaternary ammonium salts or phosphate salts, while a wide range of HBDs, including alcohols, amines, and organic acids, contribute to the diverse properties of DESs [6, 7]. Thus this chapter focuses on the formation and applications of deep eutectic solvents.

4.2 Preparation of Deep Eutectic Solvents

Water has always been regarded as an excellent solvent and aqueous solutions have been known to be utilized extensively as a type of polar solvent owing to their tremendous characteristics. Nevertheless, water-based solvents under specific conditions have drawbacks. For instance, organic reactions typically result in hydrogen evolution side reactions during the electrodeposition process and require an anhydrous environment and aqueous solvents. DESs are widely identified by their low melting point, generally, all prepared deep eutectic solvents show lower melting points as compared to their pure constituents. For example, choline chloride and urea were combined to create the eutectic mixture in a precise 1:2 molar ratio, which exhibited a melting point value of 12 °C, indicating a lower melting point of DES as compared to its pure components. The tremendously strong chemical cross-linking between hydrogen bond acceptors (HBA) and hydrogen bond donors (HBD) causes a lowering of the melting point of deep eutectics solvents. Thus, deep eutectic solvents with melting points under 50 °C have been evaluated by scientists because of their exceptional qualities, which include low cost, safe to use, and easy utilization in industrial applications. Because of their excellent biodegradable results, traditional organic solvents are increasingly replaced by DESs. Each DES components serve as both hydrogen bond donors and hydrogen bond acceptors (HBD/HBA). The formation of hydrogen bonding between HBA and HBD helps in minimizing the lattice energy of the elements themselves, hence, a deep eutectic solution generally has a lower melting temperature as compared to their single component [8]. DESs have been reported as binary or ternary mixtures, which are divided into four classes depending on the component's nature (Fig. 4.1). Type I DESs are nominated as Cat+X–zMClx, in which Cat+X– shows quaternary ammonium salt and MClx exhibits non-hydrated metal halide. However, DESs type II joined with salts, consist of hydrated metal

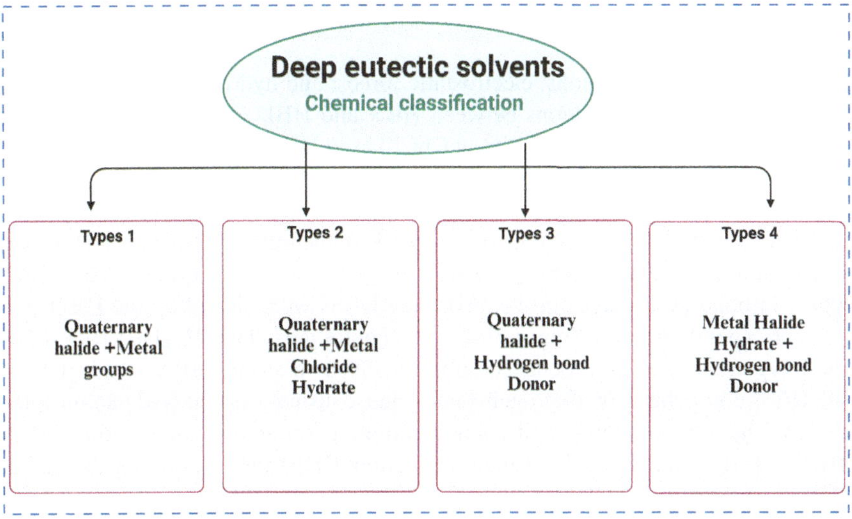

Fig. 4.1 Classification of deep eutectic solvents

halides, and generally, they are nominated as Cat+X–zMClx · yH_2O. Instead, DESs type III are mentioned as Cat+X–zRZ and have been reported to make of quaternary ammonium, solfonium or fosfonium salt, and molecular substances like alcohols, amides, and carboxylic acids. Additionally, type IV DESs have been reported in the literature to be formed by metal halides like zinc chloride ($Zncl_2$), and a molecular substance having ethylene glycol, urea, acetamide, and 1,6-hexandiol [5].

Generally, type III eutectics are reported as the most studied materials in the scientific literature, which are mostly based on choline chloride and numerous hydrogen bond donors. Because of comparatively cheap, environment-friendly as well as biodegradable choline chloride is tremendously used and has been approved "as a natural additive" for various animal species (Scientific Opinion on Safety and Efficacy of Choline Chloride as a Feed Additive for All Animal Species [9]). Type III deep eutectic solvent was initially based on choline chloride, subsequently, a plethora of constituents has been used in DES formation. In these, the HBA mainly comprises quaternary ammonium or phosphonium salts, while amides, alcohols, and carboxylic acids are most commonly used as hydrogen bond donors. Usually, type III DES is considered on of the deep eutectic solvents, however, naturally ecofriendly deep eutectic solvents are more recently acknowledged and categorized into various groups such as neutral deep eutectic solvents, meaning they contain components that can only be sugars or a combination of sugars and other different polyalcohols. Similarly, ionic liquid DES means it contains an acid and a base. "DES neutrals having bases" are types of deep eutectic solvents containing components like sugar/ polyalcohols and organic bases. However "DES neutrals having acids" are composed of sugar/polyalcohols and organic acids. Lastly, the constituents such as amino acids and sugars/organic acids are referred to as amino acids-containing natural DESs [10, 11].

Van der Waals interactions, electrostatic forces, and hydrogen bonding make up the majority of the interactions between HBA and HBD [6]. Perkins et al.'s [12] structural examination of four different ChCl-based DESs revealed that the majority of hydrogen bond interactions typically take place between the HBD and the anion (Fig. 4.2a). Some DESs can have increased HBD-HBD hydrogen bond interactions depending on the presence of functional groups in the initial components. However, HBD-anion hydrogen bond interactions usually follow them strongly. The largest hydrogen bond percentages in the next three DESs were shared by the HBDs and their corresponding anions (Fig. 4.2b, c, e) except for the [ChCl]-[U] DES (Fig. 4.2d). The oxygen atom in urea behaved substantially as a strong HBA in the [ChCl]-[U] DES, competing for hydrogen bonds and causing weaker HBD-anion interactions. The authors found similar anion-cation, anion-anion, cation-cation, HBD-HBD, and HBD-anion radial distribution function (RDF) distances among the studied DESs, specifically [ChCl]-[U], [ChCl]-[EG], and [ChCl]-[Gly]. The anion-HBD distances showed the greatest RDF changes between deep eutectic solvent components, suggesting that hydrogen bonds are the main reasons that provide neoteric characteristics to DESs [12]. Recently, new solvents in DES have been added named "hydrophobic deep eutectic solvents" which rely on hydrophobic substances like tetrabutylammonium, bromide, menthol, thymol, and fatty acids as hydrogen bond donors and long alkyl chain alcohols as HBA [13, 14]. Additionally, DESs have been prepared by active pharmaceutical constituents such as ibuprofen, lidocaine, and phenylacetic acid. However, in those cases, solvents are known as therapeutic deep eutectic solvents [15, 16].

4.3 Deep Eutectic Solvents Formation

As already mentioned, low vapor pressure and excellent thermal stability are two excellent characteristics that DESs and ionic liquids (ILs) exhibit, however, DESs are less expensive than ILs due to the lower cost and wider availability of the components, and they are also simpler to manufacture than ILs because they only require the purity of the raw materials. DESs can be made by mixing at least two components, HBD and HBA. Additionally, DESs are more environmentally friendly than ILs because of their special characteristics, which include easy structural design, excellent solubilization capacity for a variety of chemicals, non-flammability, high biodegradability, and biocompatibility, and decreased toxicity [17, 18]. DESs have been created by combining two or more substances, which produces a liquid with the unique property of having a melting point lower than that of the separate constituents. Strong hydrogen bond interactions as well as electrostatic and van der Waals interactions (although to some lesser extent), are the primary means by which DES components crosslink with one another [19]. One of the leading advantages of deep eutectic solvent is a method of preparation process because different chemicals can be used to produce DESs

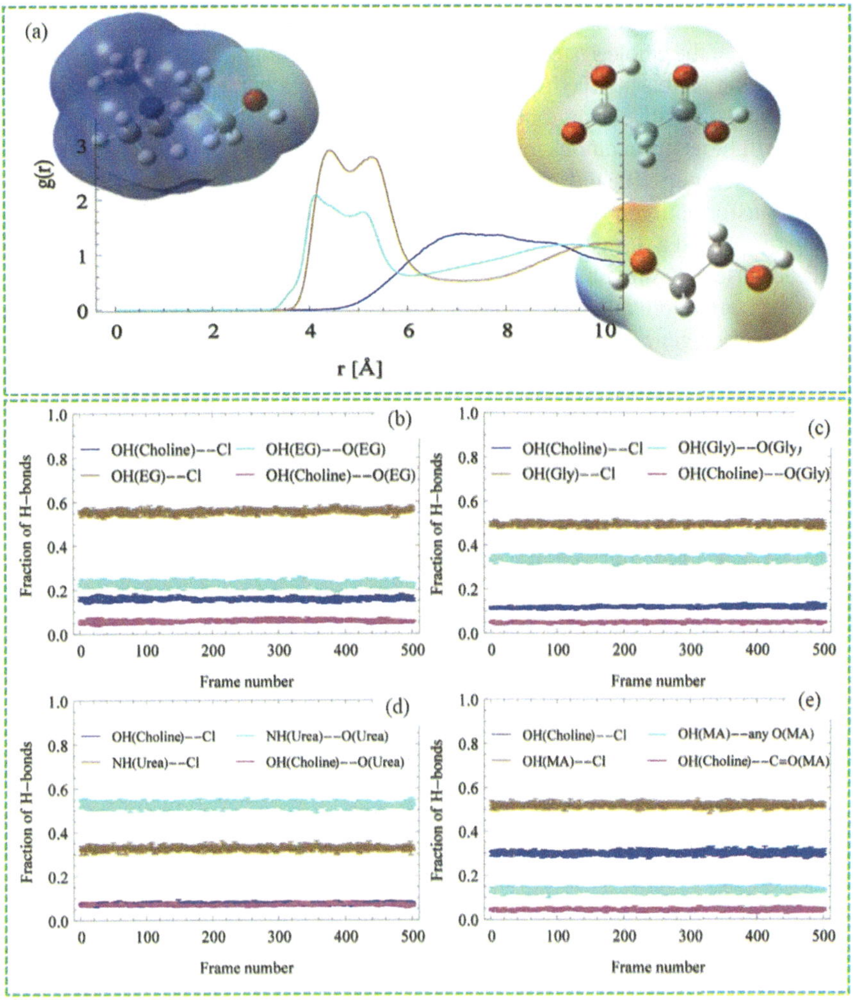

Fig. 4.2 Schematic illustration of experimental and computational studies of Choline Chloride-based deep eutectic solvents, relative hydrogen bond per frame (**a**) for ethaline (**b**), glyceline (**c**), reline (**d**), and maline (**e**). Reprinted (adapted) with permission from [12]. Copyright 2024 American Chemical Society

with a wide range of appropriate ratios in which they can be mixed, as the number of possible combinations of these solvents is huge (about 106), offering an effective tool to regulate their physical properties [20]. In terms of their synthesis procedures, deep eutectic solvents may easily be produced using three distinct techniques with varying alterations like reaction temperature or time of reaction.

It has been mentioned in the reports that the interaction of hydrogen bonding in DES is the fundamental cause behind the formation of deep eutectic solvents [21]. As a result, it has been found that the hydrogen bonding interaction between anions and hydrogen molecules prevents the system from crystallization. While it's a wide explanation of the whole system, however, it seems an ambiguous description as well. Generally, on increasing the hydrogen bonding strength between HBA and HBD, it seems that the positive charge will find extra space near cations, while the negative charge will move from anions to organic compounds. Some years ago, Mollenhauer and his team reported a study that the choline chloride and urea mixtures exhibited a minor transformation of negative charges from "chloride to urea" [22]. That's why, this study created a discussion about if the charge delocalization between HBA and HBD because of hydrogen bonding is the main reason for eutectic solvents. Later on, scientists started to explore the fundamental causes of deep eutectic solvent formation. However, generally, hydrogen bonding interactions are considered the basic reason for DES system formation. Numerous experimental techniques are being used such as infrared, differential scanning calorimetry, and nuclear magnetic resonance by the researchers to verify the eutectic system formation after carefully preparing DES [23–25] for an instant change of absorption peaks in infrared spectroscopy as well as nuclear magnetic resonance system are assumed as a clear reason for hydrogen bond interaction.

In another study, the author examined the similar cross-linking interaction of hydrogen bonding in three-dimensional distribution between both components of DES (ChCl: urea = 1:2) [26]. Similarly, chloride anions (Cl) and oxygen atoms of urea (Ourea) exhibit three-dimensional distribution along the cation core of choline. Generally, in pure materials such as salts, the combination of anionic and cationic atoms has been evaluated, however, in the case of deep eutectic solvents cationic and anionic combination is reported to be broken by the molecular compound, which is because of a contest for anions by cations and hydrogen atoms. Additionally, the increase of entropy can also be taken as an important change in the mixed system instead of only hydrogen bonding formation. Additionally, in another report, deep eutectic solvent residues were used as a value-added byproduct without any purification process. Thus, in this study urea (U) and glycidyltrimethylammonium chloride were nominated as hydrogen-bond donor and hydrogen-bond acceptor, respectively. Authors assumed that the glycidyltrimethylammonium chloride can form a deep eutectic solvent because of hydrogen bonding between chloride ion and urea [27]. They further expected that GTAC and its ring-opening product, (2,3-dihydroxypropyl) trimethylammonium chloride, possess properties that make them suitable for direct use as plant growth regulators because of exhibiting similar characteristics to widely used quaternary ammonium salts such as choline chloride and chlorocholine chloride [28, 29].

4.4 Applications of DES

Although deep eutectic solvents and conventional ionic liquids show different chemical characteristics, they are reported to exhibit similar physical properties particularly showing tunable solvent capability that can be modified to a specific type of chemistry, as well ILs and DESs exhibit a lower vapor pressure value, comparatively having wide liquid range, and non-flammable properties. However, DES has various advantages over traditionally prepared ionic liquids because they are easy to prepare, cheap, and extensively available from comparatively inexpensive components. To produce deep eutectic solvents, the two constituents are simply mixed, at a moderate temperature. Thus, it leads to producing a solvent at a lower production rate as compared to traditionally prepared ionic liquids for example imidazolium-based liquids. The lower production cost of DESs gives a reason to apply it for industrial applications at large scale. As a result of this, deep eutectic solvents have been used in several applications. Figures 4.3 and 4.4 show various types of hydrogen bond donors hydrogen bond acceptors and ternary DES preparation as well as applications of deep eutectic solvents, respectively.

Scientists are considering designing a sustainable synthetic process to prepare a recyclable solid sorbent. Deep eutectic solvents are extensively considered as an excellent and promising absorbent not only for carbon dioxide but also for Sulphur dioxide, and other gases, which is contributing excellently to environmental protection issues [30–32]. DES has been used to prepare these materials because of owing excellent efficiency and sustainability [33, 34]. There have been many studies to

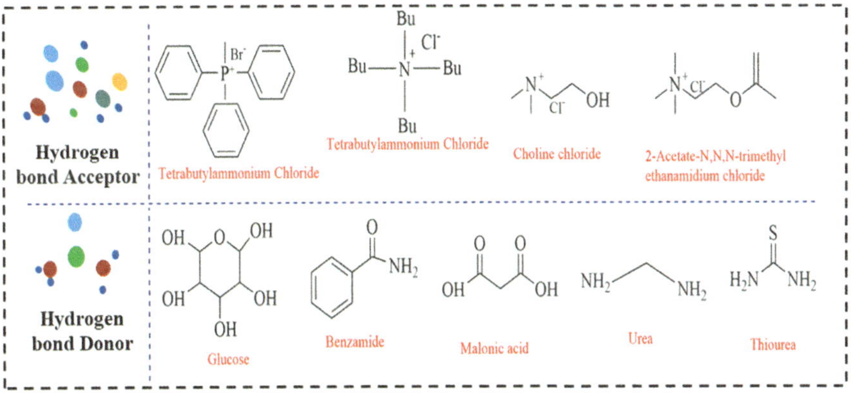

Fig. 4.3 Formation of hydrogen bond donor and hydrogen bond acceptor used in deep eutectic preparation

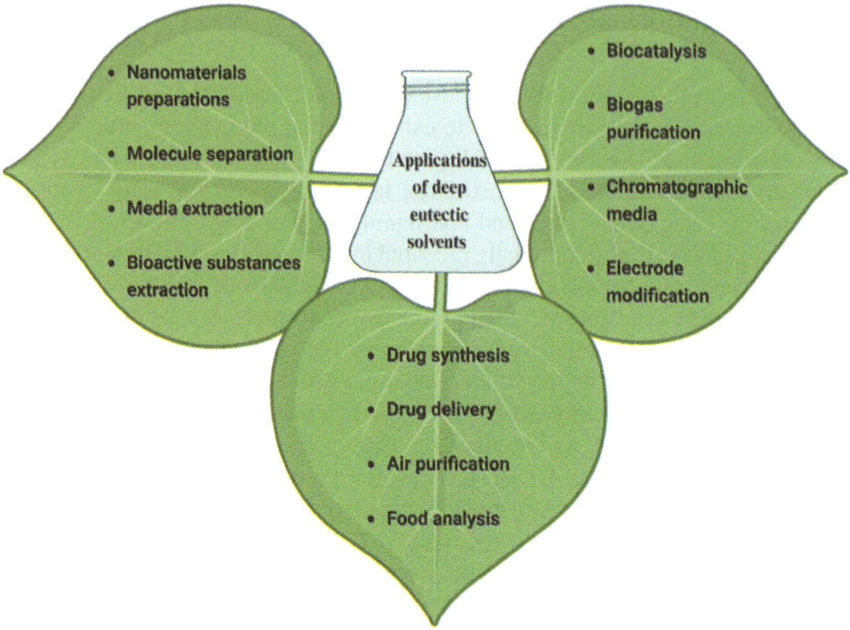

Fig. 4.4 Various applications of deep eutectic solvents

explore the synthesis and designation of successive eutectic absorbents [35]. Han et al. have proposed a method to chemically fix CO_2 to cyclic carbonates using DES containing ChCl: urea working as a catalytic agent [36].

Based on the results, it was found that this new biodegradable and "green" catalyst was extremely active and showed exceptional potential. Additionally, DESs have also been used as solvents to capture carbon dioxide [37, 38]. That's why, a study has used choline chloride and lactic acid as a solvent to capture carbon dioxide, additionally, mixtures were evaluated as being too easy to tune, however, there were much lower solubility rates for CO_2 as compared to other DES systems reported in previous studies [39]. Reduced hydrophilicity was thought to be the main reason for this phase separation. As a result of the chosen system, phase separation can be achieved successfully at 26 °C with an impressive accuracy as shown in Fig. 4.5a, b. When imidazole/ethylene glycol and olive oil combine, in a study, an emulsion was created. Once the CO_2 is bubbled into the emulsion, the emulsion will split into two transparent phases [40] (Fig. 4.5c). According to spectral investigations, the intrinsic cause of this phase shift is H_2CO_3. The initial hydrogen bond in DESs is destroyed because

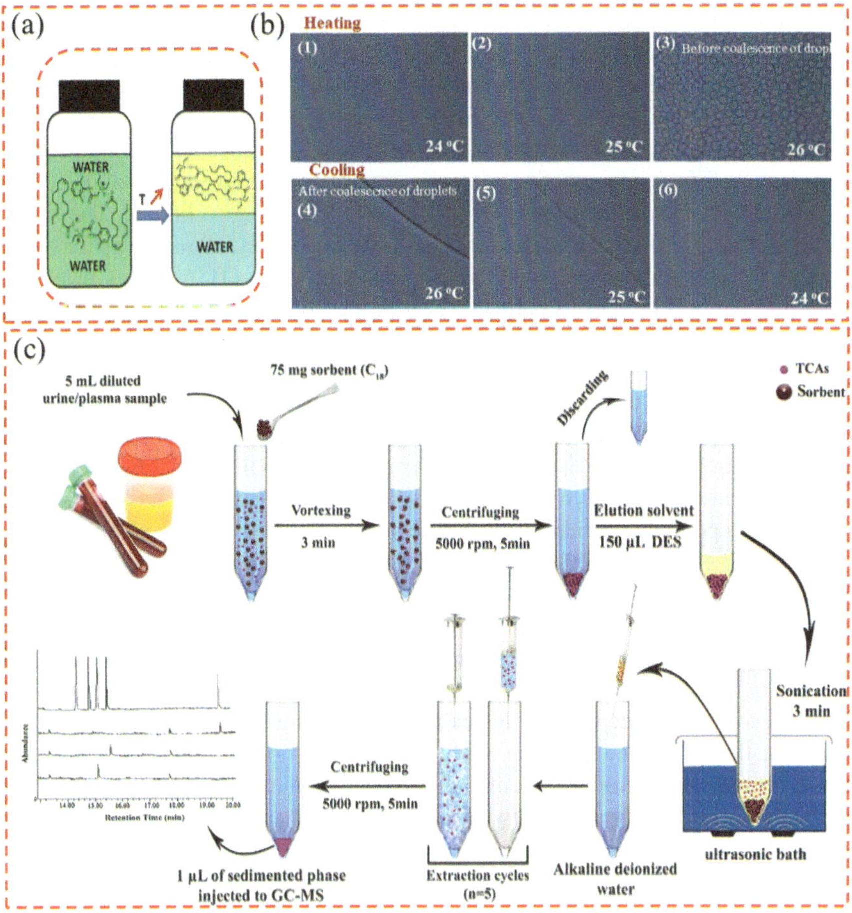

Fig. 4.5 Illustration of thermo-switchable deep eutectic solvent in water (**a**), the evolution of the liquid–liquid phase by using optical microscopy of a sample with water (50 wt.%) and DES (50 wt.%), photographs (1–3) shows heating of the sample from 24 to 26 °C, and images (4–6) are taken during cooling process (**b**) [41]. An illustration showing the proposed AA-LLME extraction procedure for tricyclic antidepressants from urine and plasma (**c**) [42]. Reproduced with permission

the acidity of H_2CO_3 is more potent than that of HBDs. This method allows for the extraction of water-soluble compounds. Similarly, Andanson et al. have suggested that a thermos-switchable DES made of oleic acid and lidocaine might be more useful [41]. It is particularly noteworthy that Mohebbi et al. [42] presented a green and sensitive sample preparation method as shown in Fig. 4.5c, termed dispersive solid phase extraction-DES based air-assisted liquid–liquid microextraction (DSPE–DES-based AALLME) in combination with gas chromatography-mass spectrometry (GC–MS). The developed method for the extraction and determination of some widely used Tricyclic antidepressants (TCAs) in urine and plasma was used. The whole adopted method is replacing the conventional and toxic method to organic solvents by using a small quantity of deep eutectic solvent about 150 μL as an elution/extraction solvent, which resulted in excellent environmentally friendly outputs such as low human health risk and low cost of routine analysis. Additionally, the proposed method is advantageous because of various findings like excellent repeatability, high enrichment factors (EFs) and extraction recoveries (ERs), low involved limit of detections (LODs), limit of quantification (LOQs), and easy operating technique. The results revealed that the proposed technique can easily be utilized in a normal way for simultaneous determination of selected tricyclic antidepressant (TCA) drugs in plasma and urine samples of various patients being treated with these drugs in the clinical pharmacology laboratories.

Medical research has been considered an important aspect in the life-saving of thousands of people, thus with the boom of technology, drug delivery has been developed and deep eutectic solvents are drawing excellent interest in potential applications and modification of drug delivery systems. Thus, DESs have been used and extensively investigated to expand these drug delivery system in medical research [43–47]. It is possible to obtain solubility as much as 22,000 times greater than water with eutectic systems including reline and malonic acid-choline chloride for several solutes including danazol, itraconazole, griseofulvin, and AMG517 [43]. Similarly, in another study, a strategy was induced for active dental composite fabrication [48]. To integrate the DES into a dental resin composite, a benzalkonium chloride and acrylic acid-based DES was prepared by using a rapid mixing technique. Their results revealed that prepared resign exhibited better mechanical properties, higher structural flexibility, and excellent antibacterial properties as compared to traditionally prepared resign composite containing benzalkonium chloride. Some of these applications are listed in the following Table 4.1.

Table 4.1 Various applications of deep eutectic solvents

HBAs	HBDs	Molar ratio of HBA: HBD	Applications and results	References
Choline chloride (ChCl)	Ethylene glycol ($C_2H_6O_2$)	1:2	A wear-resistant silver coating was made by electrolytically deposition of silver from a solution of AgCl in Deep Eutectic Solvent.	[49]
Choline chloride (ChCl)	Thiourea (CH_4N_2S)	1:2	A chemical synthesis medium for nanoparticles	[50, 51]
Choline chloride (ChCl)	Ethylene glycol ($C_2H_6O_2$)	1:2	Using a deep eutectic solvent, a novel redox flow battery has been developed	[51]
Choline chloride (ChCl)	p-Toluenesulfonic acid ($C_7H_8O_3S$)	1:1	Chemical synthesis of nanoparticles with structure-directing agents and media	[52]
Choline chloride (ChCl)	Glycerol ($C_3H_8O_3$), levulinic acid ($C_5H_8O_3$), and ethylene glycol ($C_2H_6O_2$)	1:2	There is evidence that deep eutectic solvents (DES) are excellent extraction solvents for separating azeotropic mixtures from liquids.	[53]
Citric acid ($C_6H_8O_7$)	Dimethyl urea ($C_3H_8N_2O$)	1:1.5	Nanocatalytic assembly	[54]
Choline chloride (ChCl)	Glycerol ($C_3H_8O_3$)	1:1, 1:1.5, 1:2	Among all the ratios tested, the 1:1 molar ratio resulted in the best extraction yield of glycerol from palm-oil based biodiesel	[55]
Choline chloride (ChCl)	Acrylic acid ($C_3H_4O_2$)	2.3:12.0:11.6:11.3:1	Synthesis of nanoparticles using dispersants and media	[56]
Choline chloride (ChCl)	Urea (CH_4N_2O)	1:2	CO_2 was found to be the most soluble in the tested DES when it was evaluated in terms of solubilization of CO_2, CH_4, CO, and H_2	[57]
Choline chloride (ChCl)	1,3 dimethyl urea ($C_3H_8N_2O$)	1:2	A chemical synthesis medium for nanoparticles	[58]
Choline chloride (ChCl)	Glycerol ($C_3H_8O_3$)	1:2	E-Coli bacteria are well preserved after freeze-drying using DES	[59]
Choline chloride (ChCl)	Urea (CH_4N_2O)	1:2	Electrolyte (nanostructure sensor)	[60]

(continued)

Table 4.1 (continued)

HBAs	HBDs	Molar ratio of HBA: HBD	Applications and results	References
Choline chloride (ChCl)	Ethylene glycol ($C_2H_6O_2$)	1:2	As an electrochemical solvent, DES was used to electrodeposit Zn, Sn, and Ni as well as Zn–Sn and Ni–Co alloys	[60–62]
Choline chloride (ChCl)	Urea (CH_4N_2O)	1:2	Dispersant	[63]
Choline chloride (ChCl)	Ethylene glycol ($C_2H_6O_2$)	1:2	Liquid DES dissolves gold, sulphides, and tellurides	[64]
Choline chloride (ChCl)	Ethylene glycol ($C_2H_6O_2$)	1:2	Synthesis media for nanoparticles	[65–67]
Choline chloride (ChCl)	Ethylene glycol ($C_2H_6O_2$)	1:2	In order to make $LiMnPO_4$/C nanorods for Li-ion batteries, DES was used as a solvent for synthesis	[68]
Choline chloride (ChCl)	Ethylene glycol ($C_2H_6O_2$)	1:2	Microstructure embedded with nanodroplets	[69]
Choline chloride (ChCl)	Urea (CH_4N_2O), malonic acid ($C_3H_4O_4$), and ethylene glycol ($C_2H_6O_2$)	1:2, 1:1	The deep eutectic solvents showed excellent solubility of metal oxides since protons were acting as oxygen acceptors	[70]
Choline chloride (ChCl)	Malonic acid ($C_3H_4O_4$)	1:1	Media for chemically synthesising nanoparticles with structure-directing agents	[71]
Choline chloride (ChCl)	Ethylene glycol ($C_2H_6O_2$)	1:2	An eutectic mixture of choline chloride and ethylene glycol was tested for carbon dioxide solubility	[72]
Choline chloride (ChCl)	Gallic acid ($C_7H_6O_5$), Glycerol ($C_3H_8O_3$)	1:0.25:0.25	Synthesis media for nanoparticles	[73]
Choline chloride (ChCl)	Urea (CH_4N_2O)	1:1.5, 1:2, 1:2.5	Carbon dioxide showed various solubility rate at different temperature and pressure	[33, 34]

(continued)

Table 4.1 (continued)

HBAs	HBDs	Molar ratio of HBA: HBD	Applications and results	References
Choline chloride (ChCl)	Urea (CH_4N_2O),	1:2	A ChCl/Urea DES gave excellent results for electrodeposition of Cu–In and Sm–Co and Zn, Sn and Zn/Sn alloys	[74–77]
Choline chloride (ChCl)	Ethylene glycol ($C_2H_6O_2$)	1:2	Low Coulombic (49–52%) and low energy (25–31%) efficiencies were obtained using this DES as an electrolyte in RFB	[77]
Zinc chloride ($ZnCl_2$)	Acetamide (C_2H_5NO)	Various	As anolyte for hybrid redox-flow batteries, Zn-based deep-eutectic-solvent was prepared using inexpensive and abundant materials	[78]
Choline chloride (ChCl)	Glycerol ($C_3H_8O_3$)	1:3	A dye-sensitized solar cell electrolyte based on quaternary ammonium salt-derived ionic liquid was prepared	[79]
1-Butyl-3-methylimidazolium chloride (BMIMCl)	Monoethanolamine (C_2H_7NO)	1:1, 1:2, 1:4	In an evaluation of CO_2 capture, By adjusting the deep eutectic solvent's molar ratio, CO_2 absorption can be tailored	[80]
Choline chloride (ChCl)	Chromium(III) chloride ($CrCl_3$)	1:2	Electrolyte (nanoparticle deposition)	[81]
Choline chloride (ChCl)	Ethylene glycol ($C_2H_6O_2$)	1:2	Electrolyte (nanoparticle deposition)	[62, 82, 83]
Choline chloride (ChCl)	Glycerol ($C_3H_8O_3$)	1:2	As an electrolyte medium, dye-sensitized solar cells (DSSCs) with aqueous (40% water content) choline chloride were investigated carefully.	[84]

4.5 Conclusions

Deep eutectic solvents are being utilized in various applications because of their excellent green and eco-friendly nature. The utilization of deep eutectic solvents can be considered an effective approach to enhance the performance of membranes, scaffolds, and various other composites. DESs offer several advantages as compared to other additives, such as ionic liquids, when incorporated into membrane preparation. DES utilization would contribute to reducing the environmental impact associated with the use of conventional solvents. DES-supported materials exhibit distinctive features that make them highly suitable for adsorbing various organic and inorganic guest molecules from aqueous solutions. These complexes demonstrate improved mechanochemical stability, selectivity, adsorption capacity, and permeability compared to pristine samples. Overall, the use of DESs as modifying materials holds great potential for enhancing the obtained material's performance. Continued research and development in this field can lead to the advancement of DES-supported materials for various applications.

References

1. Smith, E.L., A.P. Abbott, and K.S. Ryder. 2014. Deep eutectic solvents (DESs) and their applications. *Chemical Reviews* 114 (21): 11060–11082.
2. Płotka-Wasylka, J., et al. 2020. Deep eutectic solvents vs ionic liquids: Similarities and differences. *Microchemical Journal* 159: 105539.
3. Chen, Y., and T. Mu. 2021. Revisiting greenness of ionic liquids and deep eutectic solvents. *Green Chemical Engineering* 2 (2): 174–186.
4. Zhang, H., et al. 2022. Transitioning from ionic liquids to deep eutectic solvents. *ACS Sustainable Chemistry & Engineering* 10 (3): 1232–1245.
5. Mannu, A., et al. 2021. Promising technological and industrial applications of deep eutectic systems. *Materials* 14 (10): 2494.
6. Zhang, Q., et al. 2012. Deep eutectic solvents: Syntheses, properties and applications. *Chemical Society Reviews* 41 (21): 7108–7146.
7. Cui, Y., et al. 2017. Design, synthesis and properties of acidic deep eutectic solvents based on choline chloride. *Journal of Molecular Liquids* 236: 338–343.
8. Qin, H., et al. 2020. Overview of acidic deep eutectic solvents on synthesis, properties and applications. *Green Energy & Environment* 5 (1): 8–21.
9. Abbott, A.P., et al. 2001. Preparation of novel, moisture-stable, Lewis-acidic ionic liquids containing quaternary ammonium salts with functional side chainsElectronic supplementary information (ESI) available: Plot of conductivity vs. temperature for the ionic liquid formed from zinc chloride and choline chloride (2:1). See http://www.rsc.org/suppdata/cc/b1/b106357j. Chemical Communications (19): 2010–2011.
10. Dai, Y., J. Van Spronsen, G.-J. Witkamp, R. Verpoorte, and Y. Hae Choi. 2013. Natural deep eutectic solvents as new potential media for green technology. *Analytica Chimica Acta* 766: 61–68.
11. González, C.G., et al. 2018. Application of natural deep eutectic solvents for the "green" extraction of vanillin from vanilla pods. *Flavour and Fragrance Journal* 33 (1): 91–96.
12. Perkins, S.L., P. Painter, and C.M. Colina. 2014. Experimental and computational studies of choline chloride-based deep eutectic solvents. *Journal of Chemical & Engineering Data* 59 (11): 3652–3662.

13. Florindo, C., L.C. Branco, and I.M. Marrucho. 2019. Quest for green-solvent design: From hydrophilic to hydrophobic (deep) eutectic solvents. *Chemsuschem* 12 (8): 1549–1559.
14. van Osch, D.J., et al. 2015. Hydrophobic deep eutectic solvents as water-immiscible extractants. *Green Chemistry* 17 (9): 4518–4521.
15. Duarte, A.R.C., et al. 2017. A comparison between pure active pharmaceutical ingredients and therapeutic deep eutectic solvents: Solubility and permeability studies. *European Journal of Pharmaceutics and Biopharmaceutics* 114: 296–304.
16. Paiva, A., et al. 2014. Natural deep eutectic solvents–solvents for the 21st century. *ACS Sustainable Chemistry & Engineering* 2 (5): 1063–1071.
17. Cao, J., and E. Su. 2021. Hydrophobic deep eutectic solvents: The new generation of green solvents for diversified and colorful applications in green chemistry. *Journal of Cleaner Production* 314: 127965.
18. Khandelwal, S., Y.K. Tailor, and M. Kumar. 2016. Deep eutectic solvents (DESs) as eco-friendly and sustainable solvent/catalyst systems in organic transformations. *Journal of Molecular Liquids* 215: 345–386.
19. Wang, H., et al. 2019. Insights into the hydrogen bond interactions in deep eutectic solvents composed of choline chloride and polyols. *ACS Sustainable Chemistry & Engineering* 7 (8): 7760–7767.
20. Marcus, Y., and Y. Marcus. 2019. The variety of deep eutectic solvents. *Deep Eutectic Solvents*, 13–44.
21. Yu, D., Z. Xue, and T. Mu. 2021. Eutectics: Formation, properties, and applications. *Chemical Society Reviews* 50 (15): 8596–8638.
22. Jiang, Y., et al. 2016. Eutectic Na_2CO_3–NaCl salt: A new phase change material for high temperature thermal storage. *Solar Energy Materials and Solar Cells* 152: 155–160.
23. Chatterjee, S., et al. 2019. Hydrocarbon chain-length dependence of solvation dynamics in alcohol-based deep eutectic solvents: A two-dimensional infrared spectroscopic investigation. *The Journal of Physical Chemistry B* 123 (44): 9355–9363.
24. El Achkar, T., et al. 2020. Cyclodextrins: From solute to solvent. *Chemical Communications* 56 (23): 3385–3388.
25. Mou, H., et al. 2019. Fabricating amorphous g-C_3N_4/ZrO_2 photocatalysts by one-step pyrolysis for solar-driven ambient ammonia synthesis. *ACS Applied Materials & Interfaces* 11 (47): 44360–44365.
26. Zahn, S. 2017. Deep eutectic solvents: Similia similibus solvuntur? *Physical Chemistry Chemical Physics* 19 (5): 4041–4047.
27. Hansen, B.B., et al. 2020. Deep eutectic solvents: A review of fundamentals and applications. *Chemical Reviews* 121 (3): 1232–1285.
28. Zeng, Q., et al. 2012. Camptothecin and 10-hydroxycamptothecin accumulation in tender leaves of Camptotheca acuminata saplings after treatment with plant growth regulators. *Plant Growth Regulation* 68: 467–473.
29. Kaczmarek, D.K., et al. 2020. Transformation of indole-3-butyric acid into ionic liquids as a sustainable strategy leading to highly efficient plant growth stimulators. *ACS Sustainable Chemistry & Engineering* 8 (3): 1591–1598.
30. Jiang, B., et al. 2019. Superbase/acylamido-based deep eutectic solvents for multiple-site efficient CO_2 absorption. *Energy & Fuels* 33 (8): 7569–7577.
31. Zhang, J., et al. 2019. Role of hydrophilic ammonium-based deep eutectic solvents in SO_2 absorption. *Energy & Fuels* 34 (1): 74–81.
32. Qu, R., et al. 2018. Antioil Ag_3PO_4 nanoparticle/polydopamine/Al_2O_3 sandwich structure for complex wastewater treatment: Dynamic catalysis under natural light. *ACS Sustainable Chemistry & Engineering* 6 (6): 8019–8028.
33. Li, X., et al. 2008. Solubility of CO_2 in a choline chloride+ urea eutectic mixture. *Journal of Chemical & Engineering Data* 53 (2): 548–550.
34. Su, W.C., D.S.H. Wong, and M.H. Li. 2009. Effect of water on solubility of carbon dioxide in (aminomethanamide+ 2-hydroxy-N,N,N-trimethylethanaminium chloride). *Journal of Chemical & Engineering Data* 54 (6): 1951–1955.

35. Haider, M.B., et al. 2018. Thermodynamic and kinetic studies of CO_2 capture by glycol and amine-based deep eutectic solvents. *Journal of Chemical & Engineering Data* 63 (8): 2671–2680.
36. Zhu, A., et al. 2007. Supported choline chloride/urea as a heterogeneous catalyst for chemical fixation of carbon dioxide to cyclic carbonates. *Green Chemistry* 9 (2): 169–172.
37. Leron, R.B., and M.-H. Li. 2013. Solubility of carbon dioxide in a eutectic mixture of choline chloride and glycerol at moderate pressures. *The Journal of Chemical Thermodynamics* 57: 131–136.
38. Lin, C.-M., et al. 2014. Henry's constant of carbon dioxide-aqueous deep eutectic solvent (choline chloride/ethylene glycol, choline chloride/glycerol, choline chloride/malonic acid) systems. *The Journal of Chemical Thermodynamics* 68: 216–220.
39. Francisco, M., et al. 2013. A new low transition temperature mixture (LTTM) formed by choline chloride+ lactic acid: Characterization as solvent for CO_2 capture. *Fluid Phase Equilibria* 340: 77–84.
40. Liu, F., et al. 2021. CO 2 switchable deep eutectic solvents for reversible emulsion phase separation. *Chemical Communications* 57 (5): 627–630.
41. Longeras, O., et al. 2020. *ACS Sustainable Chemistry & Engineering* 8: 12516–12520.
42. Mohebbi, A., et al. 2018. Combination of dispersive solid phase extraction and deep eutectic solvent–based air–assisted liquid–liquid microextraction followed by gas chromatography–mass spectrometry as an efficient analytical method for the quantification of some tricyclic antidepressant drugs in biological fluids. *Journal of Chromatography A* 1571: 84–93.
43. Morrison, H.G., C.C. Sun, and S. Neervannan. 2009. Characterization of thermal behavior of deep eutectic solvents and their potential as drug solubilization vehicles. *International Journal of Pharmaceutics* 378 (1–2): 136–139.
44. Shirani, M., S. Habibollahi, and A. Akbari. 2019. Centrifuge-less deep eutectic solvent based magnetic nanofluid-linked air-agitated liquid–liquid microextraction coupled with electrothermal atomic absorption spectrometry for simultaneous determination of cadmium, lead, copper, and arsenic in food samples and non-alcoholic beverages. *Food chemistry* 281: 304–311.
45. Nam, M.W., et al. 2015. Enhanced extraction of bioactive natural products using tailor-made deep eutectic solvents: Application to flavonoid extraction from Flos sophorae. *Green Chemistry* 17 (3): 1718–1727.
46. Pradeepkumar, P., et al. 2018. Natural solvent-assisted synthesis of amphiphilic co-polymeric nanomicelles for prolonged release of camptothecin delivery. *New Journal of Chemistry* 42 (12): 10366–10375.
47. Zainal-Abidin, M.H., et al. 2019. From nanoengineering to nanomedicine: A facile route to enhance biocompatibility of graphene as a potential nano-carrier for targeted drug delivery using natural deep eutectic solvents. *Chemical Engineering Science* 195: 95–106.
48. Wang, J., et al. 2017. Incorporation of antibacterial agent derived deep eutectic solvent into an active dental composite. *Dental Materials* 33 (12): 1445–1455.
49. Abbott, A.P., et al. 2012. The electrodeposition of silver composites using deep eutectic solvents. *Physical Chemistry Chemical Physics* 14 (7): 2443–2449.
50. Mukesh, C., et al. 2014. Choline chloride–thiourea, a deep eutectic solvent for the production of chitin nanofibers. *Carbohydrate polymers* 103: 466–471.
51. Lloyd, D., T. Vainikka, and K. Kontturi. 2013. The development of an all copper hybrid redox flow battery using deep eutectic solvents. *Electrochimica Acta* 100: 18–23.
52. Gutiérrez, M.C., et al. 2011. Deep-eutectic-solvent-assisted synthesis of hierarchical carbon electrodes exhibiting capacitance retention at high current densities. *Chemistry–A European Journal* 17(38): 10533–10537.
53. Oliveira, F.S., et al. 2013. Deep eutectic solvents as extraction media for azeotropic mixtures. *Green Chemistry* 15 (5): 1326–1330.
54. Lu, J., et al. 2014. Superparamagnetic $CuFeO_2$ nanoparticles in deep eutectic solvent: An efficient and recyclable catalytic system for the synthesis of imidazo [1, 2-a] pyridines. *ChemCatChem* 6 (10): 2854–2859.

55. Hayyan, M., et al. 2010. A novel technique for separating glycerine from palm oil-based biodiesel using ionic liquids. *Fuel Processing Technology* 91 (1): 116–120.
56. Mota-Morales, J.D., et al. 2013. Synthesis of macroporous poly (acrylic acid)–carbon nanotube composites by frontal polymerization in deep-eutectic solvents. *Journal of Materials Chemistry A* 1 (12): 3970–3976.
57. Xie, Y., et al. 2016. Solubilities of CO_2, CH_4, H_2, CO and N_2 in choline chloride/urea. *Green Energy & Environment* 1 (3): 195–200.
58. Sheng, Q., R. Liu, and J. Zheng. 2012. Prussian blue nanospheres synthesized in deep eutectic solvents. *Nanoscale* 4 (21): 6880–6886.
59. Gutiérrez, M.C., et al. 2010. Bacteria incorporation in deep-eutectic solvents through freeze-drying. *Angewandte Chemie* 122 (12): 2204–2208.
60. Zheng, Y., et al. 2014. The electrochemical behavior and determination of quercetin in choline chloride/urea deep eutectic solvent electrolyte based on abrasively immobilized multi-wall carbon nanotubes modified electrode. *International Journal of Electrochemical Science* 9: 238–248.
61. You, Y., et al. 2012. Electrodeposition of Ni–Co alloys from a deep eutectic solvent. *Surface and Coatings Technology* 206 (17): 3632–3638.
62. Gu, C., and J. Tu. 2011. One-step fabrication of nanostructured Ni film with lotus effect from deep eutectic solvent. *Langmuir* 27 (16): 10132–10140.
63. Martis, P., et al. 2010. Electro-generated nickel/carbon nanotube composites in ionic liquid. *Electrochimica Acta* 55 (19): 5407–5410.
64. Jenkin, G.R., et al. 2016. The application of deep eutectic solvent ionic liquids for environmentally-friendly dissolution and recovery of precious metals. *Minerals Engineering* 87: 18–24.
65. Zhang, H., et al. 2012. Ionothermal synthesis and lithium storage performance of core/shell structured amorphous@ crystalline Ni–P nanoparticles. *CrystEngComm* 14 (23): 7942–7950.
66. Gu, C., et al. 2011. SnO_2 nanocrystallite: Novel synthetic route from deep eutectic solvent and lithium storage performance. *Functional Materials Letters* 4 (04): 377–381.
67. Gu, C., et al. 2013. One-pot synthesis of SnO_2/reduced graphene oxide nanocomposite in ionic liquid-based solution and its application for lithium ion batteries. *Materials Research Bulletin* 48 (10): 4112–4117.
68. Wu, Z., et al. 2017. Deep eutectic solvent synthesis of $LiMnPO_4$/C nanorods as a cathode material for Lithium Ion Batteries. *Materials* 10 (2): 134.
69. Gu, C.-D., and J.-P. Tu. 2011. Thermochromic behavior of chloro-nickel (II) in deep eutectic solvents and their application in thermochromic composite films. *RSC Advances* 1 (7): 1220–1227.
70. Abbott, A.P., et al. 2006. Solubility of metal oxides in deep eutectic solvents based on choline chloride. *Journal of Chemical & Engineering Data* 51 (4): 1280–1282.
71. Islam, M.S., et al. 2014. Preparation and characterization of cellulose-gelatin nanocomposite isolated from jute for biomedical application. *Indian Journal of Materials Science* 11: 105–112.
72. Leron, R.B., and M.-H. Li. 2013. Solubility of carbon dioxide in a choline chloride–ethylene glycol based deep eutectic solvent. *Thermochimica acta* 551: 14–19.
73. Shahidi, S., et al. 2015. A new X-ray contrast agent based on highly stable gum Arabic-gold nanoparticles synthesised in deep eutectic solvent. *Journal of Experimental Nanoscience* 10 (12): 911–924.
74. Abbott, A.P., et al. 2007. Electrodeposition of zinc–tin alloys from deep eutectic solvents based on choline chloride. *Journal of Electroanalytical Chemistry* 599 (2): 288–294.
75. Malaquias, J.C., et al. 2013. Electrodeposition of Cu–In alloys from a choline chloride based deep eutectic solvent for photovoltaic applications. *Electrochimica Acta* 103: 15–22.
76. Gómez, E., et al. 2011. Electrodeposition of Co, Sm and SmCo from a deep eutectic solvent. *Journal of Electroanalytical Chemistry* 658 (1–2): 18–24.
77. Bahadori, L., et al. 2016. Investigation of ammonium-and phosphonium-based deep eutectic solvents as electrolytes for a non-aqueous all-vanadium redox cell. *Journal of The Electrochemical Society* 163 (5): A632.

78. Wang, Y., et al. 2018. Zn-based eutectic mixture as anolyte for hybrid redox flow batteries. *Scientific Reports* 8 (1): 1–8.
79. Jhong, H.-R., et al. 2009. A novel deep eutectic solvent-based ionic liquid used as electrolyte for dye-sensitized solar cells. *Electrochemistry Communications* 11 (1): 209–211.
80. Cao, L., et al. 2015. Imidazole tailored deep eutectic solvents for CO_2 capture enhanced by hydrogen bonds. *Physical Chemistry Chemical Physics* 17 (41): 27306–27316.
81. Wright, A.C., et al. 2012. Nanomagnetic domains of chromium deposited on vertically-aligned carbon nanotubes. *Journal of Magnetism and Magnetic Materials* 324 (24): 4170–4174.
82. Zhang, J., et al. 2014. Enhanced corrosion resistance of Co-Sn alloy coating with a self-organized layered structure electrodeposited from deep eutectic solvent. *Journal of The Electrochemical Society* 162 (1): D1.
83. Renjith, A., A. Roy, and V. Lakshminarayanan. 2014. In situ fabrication of electrochemically grown mesoporous metallic thin films by anodic dissolution in deep eutectic solvents. *Journal of Colloid and Interface Science* 426: 270–279.
84. Boldrini, C.L., et al. 2017. Dye-sensitized solar cells that use an aqueous choline chloride-based deep eutectic solvent as effective electrolyte solution. *Energy Technology* 5 (2): 345–353.

Chapter 5
Nanocellulose Extraction by Deep Eutectic Solvents

Abstract Currently, there is a growing demand for green, eco-friendly, and cost-effective processes for nanocellulose production. Deep eutectic solvents, a type of emerging green solvent, have gained significant attention in recent years for their unique advantages in nanocellulose preparation. These advantages include non-toxicity, low cost, easy synthesis, recyclability, and biodegradability. Numerous studies have investigated the effectiveness of various techniques and chemical pretreatment for nanocellulose extraction, particularly acid hydrolysis, TEMPO oxidation, and so on in nanocellulose production. In this chapter, we provide an overview of the latest progress in utilizing these DES for nanocellulose production, focusing on the treatment procedures and key advantages they offer. Furthermore, we discuss the implementation of DES in cellulose nanofibrils and cellulose nanocrystal extraction, separately. By addressing this advancement in nanocellulose extraction by using DES, we can pave the way for the sustainable and large-scale production of nanocellulose.

5.1 Introduction

Nanocellulose, a renewable and biodegradable nanomaterial, has gained significant attention in the nanomaterial community due to its environmentally friendly properties, non-toxicity, inherent biocompatibility, and excellent physicochemical characteristics. These characteristics include low density, large specific surface area, high crystallinity, and impressive mechanical strength and modulus [1–3]. As a result, nanocellulose holds tremendous potential for a wide range of applications across various fields, including 3D printing [4], biomedicine [5], food packaging [6], functional nanopapers [7], Pickering emulsions [8], flexible electronics [9], and wearable sensors [10]. Recently, there has been a growing interest in nanocellulose applications in different fields, leading to numerous studies focusing on its preparation. Nanocellulose materials are increasingly favored in various applications due to their inherent recyclability, which significantly enhances the overall sustainability of the production process [11, 12]. The recyclability of these materials plays a pivotal role

A. Farooq et al., *Deep Eutectic Solvents in the Textile Industry*,
https://doi.org/10.1007/978-981-97-6433-4_5

in minimizing waste generation and reducing the environmental impact associated with their production.

Two main types of nanocellulose are commonly investigated: cellulose nanofibers (CNFs) and cellulose nanocrystals (CNCs). Micro/nano-sized cellulose materials, such as microcrystalline cellulose (MCC), cellulose nanofibrils (CNF), cellulose nanocrystals (CNC), and bacterial cellulose (BC), have been extensively studied and explored over the past decades [13–15]. These materials can be isolated from various sources of cellulosic material. One remarkable route for preparing CNF is through the green and sustainable isolation of CNF by TEMPO (2,2,6,6-tetramethylpiperidine-1-oxy) oxidation. TEMPO oxidation is used to oxidize cellulose nanofibers and prepared nanocellulose are TEMPO-CNF [14, 16, 17]. This method allows for the production of CNF with tailored properties. Carboxylic acid is another significant source of modifiers for CNC or CNF extraction. By using carboxylic acids, cellulose nanocrystals or nanofibers can be chemically modified to obtain desired properties [18]. Comparatively, deep eutectic solvents are cost-effective and environmentally friendly as compared to traditional solvents, thus DES is emerging as a promising approach for the preparation of nanocellulose [19]. DES shares similar characteristics with ionic liquids (IL) but is considered cheaper and less toxic. The interaction between HBA and HBD, facilitated by van der Waals forces and hydrogen bonding, leads to the formation of a transparent DES with a significantly lower melting point compared to its components [20, 21]. DES offers several advantages, including its simple synthesis, tunable structure, environmentally friendly nature, recoverability, and more [22, 23]. Furthermore, the majority of constituents used in DES systems are biodegradable, which aligns perfectly with the principles of green chemistry [24]. This highlights the potential of DES as a sustainable and eco-friendly solvent for the production of nanocellulose materials. Based on this, we sort out this chapter focuses on the extraction of cellulose nanofibrils and cellulose nanocrystals by using deep eutectic solvents.

5.2 DES for Cellulose Nanofibrils Extraction

Most recently, deep eutectic solvents have been utilized in biorefined industries to fabricate nanocellulose. This breakthrough is successfully used in research to extract nanocellulose, results proving it an alternative green route to prepare non-derivatized and anionic nanocellulose [25–28]. Thus, Choline chloride and urea-based deep eutectic solvent with a molar ratio 1:2 was prepared in a study to prepare non-hydrolytic pre-treatment media to prepare cellulose nanofibrils from birch cellulose pulp using mechanical disintegration of microfluidizer. This study acknowledged the DES treatment conducted at about 100 °C and then samples were washed with deionized water to eliminate DES molecules. Later on, mechanical disintegration at three different levels of microfluidizer was used to obtain CNF and their effects on obtained nano fiber's properties were compared. Cellulose nanofibrils have been prepared with an individual width containing 2–5 nm, however, nanofibrils bundle width ranged from 15 to 200 nm by using a deep eutectic solvent containing choline chloride and urea. Additionally, wide-angle X-ray diffraction and degree of polymerization revealed that both the cellulose crystalline structure and degree of polymerization of cellulose samples remained the same after DES pre-treatment. Similarly, a gel-like cellulose nanofibrils structure was produced, with a mean equivalent spherical diameter of 262 nm was successfully achieved by using two types of deep eutectic solvents containing choline chloride-urea and choline chloride-oxalic acid from ramie fibers [29]. For CNF extraction, most of the studies have been reported to use urea-based deep eutectic solvents for example in a study [27], cellulose nanofibrils production was done by preparing first of all deep eutectic systems having urea and ammonium thiocyanate or guanidine hydrochloride at 100 °C and then cellulose fiber samples (derived from wood) were pass through this system to prepare cellulose nanofibrils. Results of this study reveal that fibrillated nanofibers were in the width range of about 13.0–19.3 nm prepared from both types of DES systems. Additionally, the DES system has been proved to swell the cellulose structure as results exhibited in the enhancement of fibers' lateral dimensions. On the evaluation of cellulose nanofibrils properties, authors revealed that both of the adopted systems of DES had excellently assisted the cellulose fibers to disintegrate into nanofibrils during the mechanical disintegration process. TEM images revealed that large nanofibril bundles having a width of around 100 nm were obtained in the samples treated with the first level of the mechanical disintegration process for nano-fibrillation. However, well-dispersed and comparatively small nanofibril bundles were observed. Additionally, some large fragment dimensions of fiber cell walls were also observed with all three levels of mechanical disintegration. Later on, cellulose nanofibril film was prepared and the properties of the composite were evaluated deeply as all prepared (CNF) films obtained excellent tensile strength ranging from approximately 135–189 MPa that were noted very close to periodate-based nanocellulose films having tensile strength of 130–163 MPa [30]. In this study, cellulose fibers were evenly disintegrated into nanofibers through mechanical disintegration, however, some of the unexpected white flakes were observed on cellulose nanofibril films. However, excellent thermal properties of CNF films were obtained when the samples were evaluated in air and N_2 atmosphere conditions, however, in N_2 atmosphere, cellulose nanofibrils film samples showed comparatively similar behavior but with monomodal degradation curves compared to air.

After one year of this study, the authors published their report on preparing cationic nanocellulose [31], thus under mild reaction conditions (70 °C, 10 min) dialdehyde cellulose samples were cationized into cationic dialdehyde celluloses by using the deep eutectic solvent system. Subsequently, cationized cellulose fibers were converted into nanocellulose by using mechanical disintegration techniques. Thus, according to the reaction conditions (initial aldehyde content), the morphology of nanocellulose was tailored to prepare highly cation cellulose nanofibrils or cellulose nanocrystals. Results of the study revealed that the average widths of obtained cellulose nanofibrils and cellulose nanocrystals were 4.6 ± 1.1 nm and 5.7 ± 1.3 nm, respectively. Various cellulose board grades (waste board, fluting, and waste milk container board) were used in a study to obtain cellulose nanofibrils by pretreating with a deep eutectic solvent containing choline chloride-urea and mechanically disintegrated by disc grinding. In this study, DES's obtained cellulose nanofibrils properties were carefully compared with those obtained through micro-fluidization. Results revealed that overall, choline chloride-urea-based DES pretreatment had excellently enhanced the nanofibrillation of cellulose pulp, moreover, DES pretreated obtained cellulose nanofibrils suspension was showing viscous and gel-like appearance exhibiting shear thinning behavior. Additionally, cellulose nanofibrils were noted to maintain their initial crystalline structure with a value of crystallinity index of 47–61% [32].

Highly charged cellulose nanofibers were produced in a study from wood cellulose pulp with the help of a reactive deep eutectic solvent containing sulfamic acid and urea. At first, DES was prepared at 80 °C with different molar ratios of sulfamic acid and urea (1:4, 1:3, or 1:2). Later on, cellulose fibers were passed through DES at 80 °C, followed by heating at 150 °C for a specific reaction time (90 min). Obtained samples were named sulfated cellulose, which was easily disintegrated into sulfated cellulose nanofibrils. Because of their excellent high viscosity even at low concentrations sulfated cellulose nanofibrils were noted to show tremendous properties to be utilized as a rheology modifier such as in hygienic or food applications [33]. Afterward, an economical green, and facile method was used to extract cellulose nanofibrils by using choline chloride and lactic acid-based DES from moso bamboo [34]. Authors have reported this method as an excellent procedure that can significantly reduce environmental issues as well as the cost of energy and liquid wastage during pulping and bleaching process [35]. The results of this report noted that lignin was sufficiently eliminated and a higher percentage of cellulose (91%) was obtained on passing moso bamboo through choline chloride and the lactic acid at specific reaction conditions (120 °C, 3 h, and solid to liquid ratio 1:25). Later on, cellulose nanofibrils were prepared by a simple mechanical technique and the width of the fibers was noted between 20 and 80 nm. Additionally, obtained cellulose nanofibrils were reported to be used to prepare strong films that can feature 163–213 MPa tensile strength.

Wood cellulose fibers and microcrystalline cellulose have been used to produce regenerated cellulose nanofibrils by mechanical disintegration of cellulose fibers after dissolution from guanidine hydrochloride and anhydrous phosphoric acid (molar ratio 1:2) based DES at room temperature (25 °C) [36]. During the dissolution process, the degree of cellulose sample was observed to decrease, however, the crystallinity of the samples altered from cellulose I to cellulose II. After the precise reaction conditions (precipitation and washing) cellulose fibers were disintegrated into a uniform nanosized material having a diameter of about 6 nm. Later on, cellulose nanofibrils were investigated to be used in poly(vinyl alcohol) composite films as fillers because of their excellent small and fibrous as well as flexible nature cellulose nanofibrils were proven to enhance the elongation properties of poly(vinyl alcohol) films even at lower CNF concentration (1–5 wt.%). The results of this study exhibited an excellent remedy for preparing cellulose nanofibrils showing uniform size distribution and controllable reinforcement characteristics. Deep eutectic solvents have been considered a fairly new class of green solvents applied in wide areas. Thus, in another study [27], CNF production was done from wood-cellulose fibers by preparing deep eutectic solvent at 100 °C with components urea ammonium thiocyanate or guanidine hydrochloride and later on, treating cellulose fibers with DES to disintegrate properly. In the end, mechanical disintegration with a microfluidizer was done at three different levels. The characteristics of obtained cellulose nanofibrils were examined and subsequently used for the preparation of nanofibril films. DES systems were considered to be able to easily loosen and swelling of cellulose fiber structure as lateral dimensions of the fibers were enhanced. Authors successfully prepared CNF with great dimensions such as a width ranging from 13.0 to 19.3 nm and the prepared nanofibril film was thermally stable with excellent mechanical properties such as high tensile strengths (135–189 MPa) and elastic modulus (6.4–7.7 GPa). DESs are green chemicals that have excellent potential to replace traditional solvents easily in various chemical reactions. Thus, urea and lithium chloride-based DES were prepared as a reaction medium in the anionic functionalization of wood cellulose with succinic anhydride to isolate cellulose nanofibrils [37]. Authors successfully prepared CNF and during this whole process effects of reaction temperature and time were assessed on the yield and carboxylic content of cellulose nanofibrils. The results showed that three samples with the highest carboxyl contents showed excellent dimensions (diameter 2–7 nm) as examined by atomic force microscopy (AFM). Similarly, triethylmethylammonium chloride and imidazole-based deep eutectic solvents have been used in a study as a reaction medium for successful cationization of cellulose samples with trimethylglycine (betaine) hydrochloride in the presence of p-Toluenesulfonyl (tosyl) chloride [38]. Later on, a mild mechanical disintegration was done of cationized cellulose. Cationic cellulose nanofibers were produced carefully from wood pulp exhibiting 4.7 ± 2.0 nm fiber diameter. Energy cane bagasse has been used to prepare cellulose nanofibers with the help of microwave-assisted deep eutectic solvent treatment combined with ultrasonication [39] The schematic diagram has been shown in Fig. 5.1a–c. Authors have reported that with the addition of lignin-containing cellulose nanofibrils, the stability of polyanionic cellulose (PAC) film-forming suspension was improved which was evidenced by zeta potential (Fig. 5.1d) and viscosity values. Similarly, the authors acknowledged that the lignin-containing CNF composite films combined with polyanionic cellulose exhibited excellent tunable mechanical and UV-resistant properties (depending on the amount and type of CNF added). Such as composite film containing 5% lignin-containing CNF exhibited excellent mechanical properties (tensile strength 55.8 MPa, strain to break 26.3%) as shown in Fig. 5.1e and excellent ultraviolet protection capability of about 2.9% ultraviolet transmittance. Lignin-containing cellulose nanofibril (0.5 wt.%) prepared by ultrasonication at 960 W for 30 min (was named D110) as shown in Fig. 5.1f shows excellent stability and can stand for about one month. The resultant lignin-contained cellulose nanofibers exhibited a high entangled network because of the presence of lignin between cellulose nanofibers as shown in Fig. 5.1f, g. It shows that lignin and cellulose nanofibrils were present in the suspension of D110, resulting in a highly entangled network structure. Additionally in this study, cellulose nanofibrils suspension containing 5.88 ± 0.42% lignin was prepared for comparison by similar ultrasonic treatment from cellulose samples (isolated from energy cane bagasse) by using sodium hydroxide (20 wt.%)/sodium chlorite treatment and named as Rs20. TEM and digital photographs

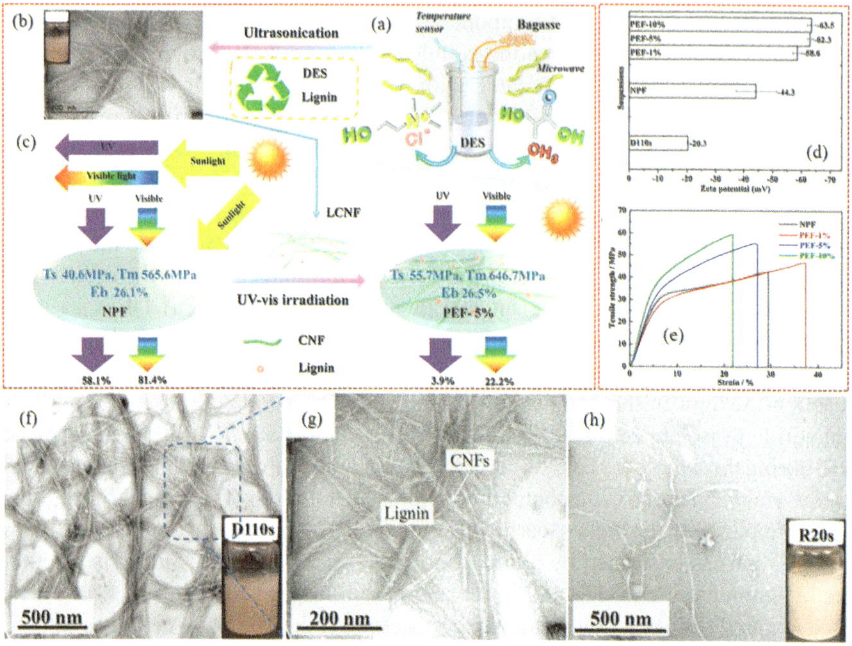

Fig. 5.1 Schematic illustration of preparation lignin containing cellulose nanofibrils and composite films, microwave reaction system (**a**), TEM and digital images of LCNF D110 (**b**), prepared film UV-resistant (c), zeta potential of CNF solution (d) and mechanical (e) properties of composite film and composite film having lignin containing cellulose nanofibrils (5%) from Energy cane bagasse. TEM and digital images of D110 (**f**, **g**) and Rs20 (**h**). However, TS, TM, and Eb show tensile strength, tensile modulus, and elongation at break, respectively [39]. Reproduced with permission

of Rs20 have been shown in Fig. 5.1h, on comparison authors have mentioned that the structure morphology of D110 exhibited a more entangled structure as compared to Rs20 because of the presence of lignin in D110 samples helping in binding and facilitating the formation of strong network structure [40].

5.3 Cellulose Nanocrystals Extraction

The most abundant natural polymer in the world is cellulose which is a renewable, biodegradable, and nontoxic material. Approximately 10^{11} to 10^{12} tons of cellulose are produced in a photosynthesis process each year. It is usually found in the cell walls of woody plants where it is combined with other polysaccharides (the so-called hemicelluloses) such as lignin and cellulose. It is widely used in a variety of industries, such as film, fiber, paper, pharmaceuticals, and paint [41]. As cellulose is a biodegradable, nontoxic, and renewable material, however, its dissolution in water and most other common organic solvents has been reported as not so good, thus the chemical processing of cellulose is still challenging. Generally, it is essential to eliminate inter- and intramolecular hydrogen bonds between cellulose molecules for them to dissolve [42]. To minimize this issue, ionic liquids as well as molten salt hydrates have been to dissolve completely the cellulose molecules [42–45]. Nevertheless, ionic liquids have been considered expensive, not easy to prepare, as well as have environmental issues. That's why biobased systems have been used to extract nanocellulose by using deep eutectic solvents in various studies [46]. Similarly, DES solvent has been utilized in the extraction of nanocellulosic materials such as nanocellulose crystals and nanocellulose fibers. Figure 5.2a shows the reaction mechanism of deep eutectic solvents and lignin carbohydrate complexes. Choline chloride and urea having a molar ratio of 1:2 have also been evaluated in cellulose fiber treatment, results revealed that the excellent gel-like material (cellulose nanofibrils) was obtained by wood pulp after treating with deep eutectic solvents. DES-based integration has been used for the destruction of non-cellulosic structural integrity without damaging the chemical composition of the cellulosic part by an excellent swelling and decaying hydrogen bond between chain structural units of wood pulp fibers [27]. Schematic illustration of nanocellulose process and resulting nanocellulose materials are presented in Fig. 5.2b.

Deep eutectic solvents have been used for hydrolysis and dissolution of amorphous domains of cellulosic fiber to obtain cellulose nanocrystals. Thus, in 2016 Sirviö et al. [49] successfully explored four types of different DES systems based on choline chloride and organic acids to dissociate cellulose pulp into CNC using a temperature between 60 and 120 °C and reaction time set between 2 and 6 h. ChCl-oxalic acid was evaluated successful DES system among all four DESs tested to isolate CNC from cellulose pulp. Results of this study show that choline chloride and oxalic acid based DES produced CNCs exhibited dimensions as width in the range of 10–16 nm and length was examined as 337–390 nm with carboxylic acid content as 0.20–0.27 mmol/g. Recently, nano cellulose extraction by using deep eutectic solvents has received excellent attention in numerous research directions because of their tremendous properties and advantages including facile fabrication, low melting temperature, and tunable physicochemical properties [50, 51]. Thus, in a recent study, cellulose nanocrystals were prepared by using deep eutectic solvents from wood cellulose fibers. Choline chloride and organic acids including oxalic acid p-toluenesulfonic acid monohydrate, and levulinic acid were carefully utilized to isolate CNC. In this study, deep eutectic solvents were prepared at elevated temperatures ranging between 60 and 100 °C by combining choline chloride and organic acids. Subsequently, less ordered amorphous regions of wood cellulose fibers were hydrolyzed. Results of this study reveal that all the DES treatments showed excellent degradation of wood fibers into micro cellulose fibers. Later on, CNCs were reported to be isolated after mechanical disintegration [49]. Moreover, the results of this study reveal more outstanding findings that obtained nano cellulose crystals

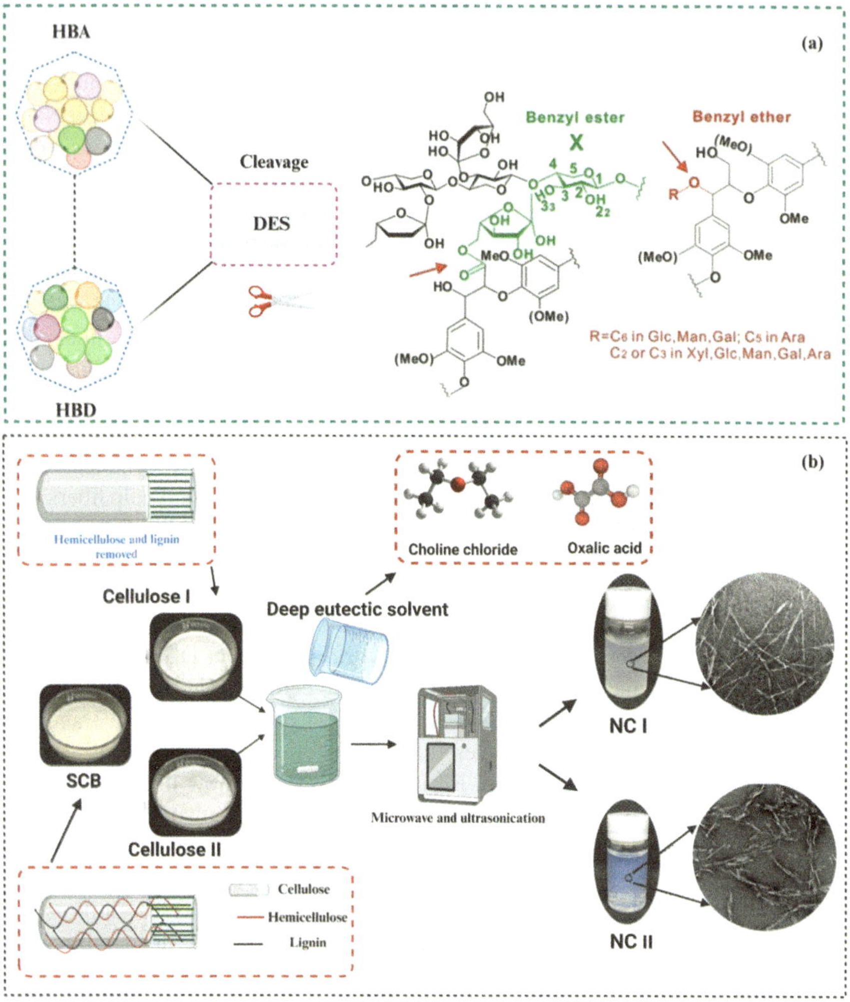

Fig. 5.2 Illustration of reaction between lignin carbohydrate complexes and deep eutectic solvents (**a**) which has been modified with the permission of [47], illustration showing the preparation of nanocellulose-I and nanocellulose-II using natural deep eutectic solvents (NDES) in conjunction with microwave and ultrasound technology (**b**) [48]. Reproduced with permission

exhibited distinct rod-like crystal structures owing to tremendous crystallinity and thermal properties as well as being subjected to excellently dispersed in aqueous solution. Cellulose nanocrystals were produced in another study yielding up to 74.2% CNC through 3 min of microwave-assisted DES pretreatment and subsequently high-intensity ultra-sonication process was used as shown in Fig. 5.3a [52]. The obtained results revealed excellent CNC properties including 3–25 nm diameter and length between 100 and 350 nm. Figure 5.3b–d shows optical and polarized images of CNC suspension at various temperatures. The isolated cellulose nanocrystal samples from cotton fibers provide an efficient and green method by using deep eutectic solvents containing choline chloride and oxalic acid, which was expected to improve the refinery utilization of cotton feedstock. Similarly, a green and reusable technique was invented to prepare CNC by using a $FeCl_3$-catalyzed deep eutectic solvent system as shown in Fig. 5.3e. $FeCl_3$-catalyzed deep eutectic solvent was composed of oxalic acid dehydrate, choline chloride, and $FeCl_3 \cdot 6H_2O$ in a molar ratio of 4.43:1:0.1. Additionally, results revealed that the obtained cellulose nanocrystals were within a diameter range of 5–20 nm and length of 50–300 nm when analyzed and examined by various techniques Fig. 5.3f, g. Additionally, cellulose nanocrystals extracted from bleached eucalyptus kraft pulp were reported to higher yield percentage prepared by a one-step $FeCl_3$-catalyzed deep eutectic solvent system

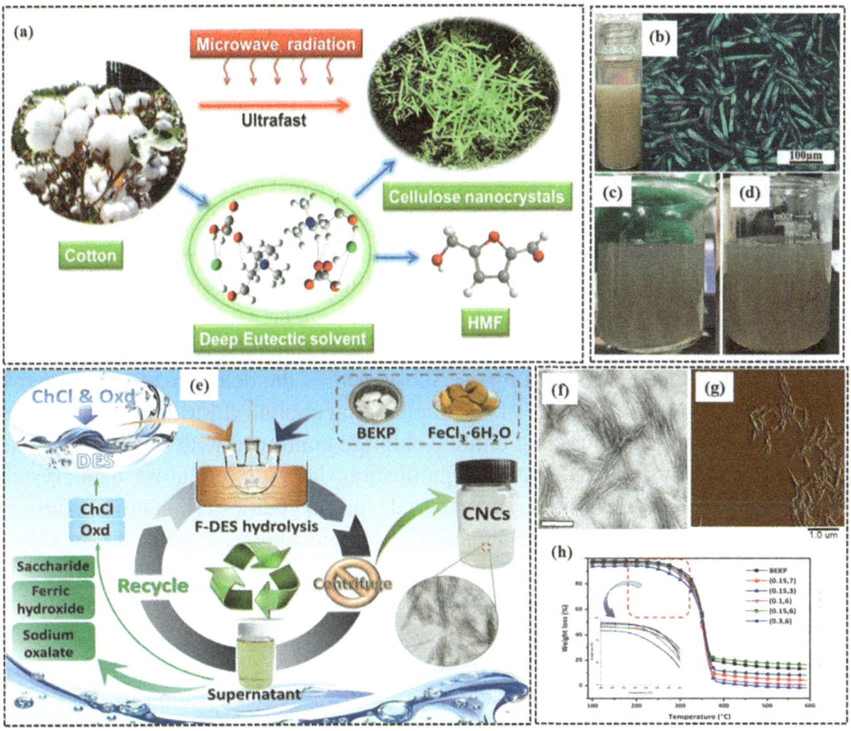

Fig. 5.3 **a** Schematic diagram for isolating CNCs from cotton by microwave-assisted DES pretreatment and subsequent ultra-sonication [52]. **b** Optical image and polarized optical microscope photo of microwave-DES pretreated CNC suspensions at 80 °C. **c**, **d** Polarized optical microscope photograph microwave-DES pretreated CNC suspension at 90 °C and 100 °C, respectively. **e** Preparation of thermally stable and dispersible cellulose nanocrystals by using recyclable $FeCl_3$-catalyzed deep eutectic solvents system [54]. **f** TEM image of obtained cellulose nanocrystals. **g** Shows the AFM images of CNC samples. **h** TGA analysis of CNC prepared by $FeCl_3$-catalyzed DES system. Reproduced with permission

under mild reaction conditions (80 °C, 6 h). Moreover, higher thermal behavior of the samples was reported as the onset thermal degradation temperature was over 310 °C (Fig. 5.3h) showing higher thermal values than the traditional sulfuric acid hydrolyzed ones. Considering cellulose the most abundant natural polymeric source and promising green material Laitinen et al. fabricated CNC in deep eutectic solvents and successfully used it as a novel biochemical for aqueous Pickering emulsification of marine diesel oil [26]. Results of this study reveal that CNC samples were exhibiting individual rod-like structures showing a diameter of 3–8 nm and length of 50–350 nm. The performance of prepared DES–CNCs was comparable to that of commercial CNCs exhibiting effectiveness at 0.1% dispersant dosage because of owning carboxyl content of 0.3 mmol/g and higher crystallinity (68%) of cellulose nanocrystals making it promising for practical application in oil–water Pickering emulsion stabilizers. Later on, similar findings were made to prepare cellulose nanocrystals from various types of raw materials by using deep eutectic solvents [53]. The results of this study prove that the use of DES as a catalyst and solvent introduced a green chemical process that didn't create any harmful waste during the experiment as well as it was considered economical because of its recyclability (>85%). Additionally, the resulting cellulose nanocrystals show excellent dimensions (diameter 50–100 nm, length 500–800 nm) which were successfully assembled by a simple technique by treating various biomass feedstocks with deep eutectic solvents and subsequent high-pressure homogenization was done.

Robinia pseudoacacia wood was passed through microwave-assisted autohydrolysis and deep eutectic solvents to prepare multiproduct bio-refinery to selectively isolate the main components of Robinia pseudoacacia wood including cellulose nanocrystals. Microwave-assisted autohydrolysis enabled the recovery of hemicellulose (76%) in the form of oligomers. Subsequently, the delignification of the raw material was done with the help of DES containing choline chloride and lactic acid. Deep eutectic solvent containing choline chloride combined with oxalic acid treatment in this study revealed 86% delignification, later on, cellulose nanocrystals were obtained from cellulosic-rich material via enzymatic-fermentative pathway and obtained cellulose nanocrystals showing dimensions as 27–550 nm in length and 2–12 nm in [55]. Recently, microcrystalline cellulose has been converted into cellulose nanocrystals by chemical hydrated DES combined with mechanical high shear force methods [56]. Results revealed that with the increase of time, the yield of cellulose nanocrystals was also increased showing yield percentages as 27.2% and 65.2% for 0.5 and 2.5 h, respectively. Meanwhile, the duration of DES treatment did not show significant changes in the dimensions and morphology of the obtained CNC. The author reported this study of using a deep eutectic solvent with a mechanical high shear force method as more eco-friendly, low cost, and higher yield as compared to sulfuric acid hydrolysis for CNC extraction from microcrystalline cellulose. Similarly, acidic deep eutectic solvents containing choline chloride

Fig. 5.4 Nanocellulose production by using hydrated deep eutectic solvent combined with ultrasonic treatment (**a**), TEM images, and diameter distribution histograms of 10% deep eutectic solvent (DES)-pretreated cellulose nanofibers (**b**), 30% DES-pretreated CNCs (**c**) [3]. Schematic illustration for preparing lignin-cellulose nanocrystals (**d**), TEM images and width distribution (**e**), and AFM image and height distribution profile of lignin-containing cellulose nanocrystals (**f**) [59]

and oxalic acid dehydrate (molar ratio of 1:1) have been used to prepare cellulose nanocrystals. The authors successfully compared and analyzed the experiment design by changing cotton fiber concentration, temperature, and reaction time with the help of yield and crystallinity of cellulose nanocrystals [57]. In a study, cellulose from wood fibers was degraded to cellulose nanocrystals by using a eutectic mixture Additionally, cellulose nanocrystals with excellent properties such as higher crystallinity (~82%), excellent thermal stability, and uniform morphological dimensions (diameter 3–25 nm, length 100–350 nm) [58].

Kraft pulp pretreatment with deep eutectic solvent has been done to mass-prepare nanocellulose through a new approach by using choline chloride/oxalic acid combined with the ultrasonic process. This study acknowledged that one pass of lignocellulosic material (kraft pulp) through hydrated DESs at 80 °C for 1 h was enough to isolate cellulose nanofiber or cellulose nanocrystals from kraft pulp using an 800 W ultrasonic treatment [3]. Moreover, 10% DES-pretreated kraft pulp was dissociated into cellulose nanofibrils when 800 W ultrasonic treatment was done for 20 min, and cellulose nanocrystals were obtained when kraft pulp was treated with 20% or 30% DES using the same ultrasonic treatment (Fig. 5.4a). Additionally, resultant CNFs and CNCs exhibited nano-order scale dimensions (Fig. 5.4b, c), higher crystallinity values, and excellent thermal stability behavior. Similarly, a recent study demonstrated that promising solvent DES can be utilized in both lignocellulosic material utilization and lignin-containing cellulose nanocrystal extraction [59] as shown in Fig. 5.4d. In this study, lignin-containing cellulose nanocrystals have been prepared by treating undervalued thermomechanical pulp both with binary deep eutectic solvents (choline chloride and oxalic acid) and ternary DES containing choline chloride, oxalic acid, and p-toluenesulfonic acid, subsequently, mechanical disintegration was accomplished. As shown in Fig. 5.4e, f results of the study exhibited that the lignin-containing cellulose nanocrystals were excellent in dimensions (diameter ~6 nm and thickness ~3.3 nm) by using ternary DES with specified reaction conditions (80 °C, 3 h).

The effect of pretreatment conditions on the preparation of cellulose nanocrystals has also been studied by Ling et al. [60]. Thus, DES components containing choline chloride and oxalic acid with different molar ratios of 1:1, 1:2, and 1:3 were prepared. Temperature variation was done by maintaining reaction conditions at 80 and 100 °C. Subsequently, ultra-sonication was done to obtain the cellulose nanocrystals. In this study, results exhibited that with the increase of the amount of oxalic acid the carboxylate content in cellulose nanocrystals was increased that were gained by O–H esterification of cellulose at the C6 position. Thus, resulting in hydrogen bond breakage and excellently enhancing the yield of cellulose nanocrystals. Table 5.1 shows various applications of deep eutectic solvents for nanocellulose extraction, reaction conditions, and biomass sources.

Table 5.1 Biomass used in cellulose nanocrystals preparation, reaction conditions (time, temperature, DESs and their molar ratio), CNC properties

DES components		Molar ratio		Biomass sources	Reaction conditions			Product obtained	Properties			References
HBA	HBD	HBA	HBD		Time	Temperature (°C)	Mechanical treatments		Dimensions	Yield (%)	Crystallinity	
ChCl	Oxalic acid	1	1	Cotton fibers	3 min	80	Ultra-sonication	CNC	L(3–25 nm) W(100–350 nm)	74.2	65–82%	[52]
ChCl	Oxalic acid	1	1	Cotton fibers	6 h	90	High pressure homogenization	CNC	D(50–100) L(50–800)	55.1	63.5–77.6%	[53]
ChCl	Oxalic acid	1	1	Cotton fibers	6 h	95	None	CNC	H(11 nm) L(257 nm)	43.6 ± 1.9	81%	[57]
ChCl	Farmic acid	1	2	Hardwood bleached kraft pulp	3 h	100	Mechanical extrusion and colloid milling	CNF	W(6 nm)	72–88	79.3%	[61]
ChCl	Farmic acid	1	2	Microcrystalline cellulose	2 h	90	Ball milling	CNC and CNF	W(8.3–10.6 nm)	95	72.07%	[62]
ChCl-	Oxalic acid	1	1	Bleached rice straw pulp	4 h	80	None	CNC	D (8 nm) L(302 nm)	55.1	76.7%	[63]
ChCl	Urea	1	2	Microcrystalline cellulose	2 h, 2 days, 4 days	110	None	CNC	Aver. size of 20 × 100 × 700 nm	//	79–85%	[64]
ChCl	Tartaric acid	1	2	Moso bamboo	0.16 h	160	microwave irradiation	CNF	W (9.62 nm)	//	75%	[65]
ChCl	Levulinic acid	1	2	Wheat, corn, and rapeseed	16 h	100	Micro-fluidizatio	CNF	W(3–12 nm)	//	46–49%	[66]
ChCl	Maleic acid	1	1	Cellulose	3 h	100	Colloid milling	CNF	45	//	//	[67]

(continued)

Table 5.1 (continued)

DES components		Molar ratio		Biomass sources	Reaction conditions			Product obtained	Properties			References
HBA	HBD	HBA	HBD		Time	Temperature (°C)	Mechanical treatments		Dimensions	Yield (%)	Crystallinity	
ChCl	Oxalic acid	1	1	Softwood	6 h	90	Mechanical blending	Lignin containing CNC	W(7 nm) Thickness (4 nm)	57	60%	[68]
	Oxalic acid dihydrate	1		Softwood	30 min	110	Microfluidization	Lignin containing CNC	W (7 nm) H (4 nm) L (200–300 nm)	70	55–72%	[69]
ChCl	Oxalic acid (1:1)	1	1	Softwood	2–6 h	100–120	Microfluidization	CNC	W (9–17 nm L (310–410 nm)	//	66–71	[49]
ChCl	Acetic acid	1	2	Hardwood bleached kraft pulp	3 h	50–100	Mechanical extrusion and colloid milling	CNF	W(15–30)	72–88	73.2–79.6%	[61]

5.4 Conclusion

In conclusion, this chapter presents a comprehensive overview of the recent advancements in the utilization of acidic deep eutectic solvent systems based on different HBA and HBD components for the preparation of nanocellulose. In comparison to traditional nanocellulose preparation techniques, the DES treatment demonstrates excellent potential in terms of cost-effectiveness, recyclability of the solvent, reduced energy and water consumption, high production yield, as well as improved thermostability, dispersibility, and surface carboxylation of the resulting nanocellulose, especially by using ChCl/carboxylic acid DES treatment. To further enhance the utilization of DES for large-scale nanocellulose production, future efforts should focus on enhancing the efficiency of the DES system for fabricating high-content nanocellulose, conducting cost recycling techniques, and performing life cycle assessments to evaluate the overall process.

References

1. Gan, P., et al. 2020. Thermal properties of nanocellulose-reinforced composites: A review. *Journal of Applied Polymer Science* 137 (11): 48544.
2. Omran, A.A.B., et al. 2021. Micro-and nanocellulose in polymer composite materials: A review. *Polymers* 13 (2): 231.
3. Ma, Y., et al. 2019. Production of nanocellulose using hydrated deep eutectic solvent combined with ultrasonic treatment. *ACS Omega* 4 (5): 8539–8547.
4. Sultan, S., and A.P. Mathew. 2018. 3D printed scaffolds with gradient porosity based on a cellulose nanocrystal hydrogel. *Nanoscale* 10 (9): 4421–4431.
5. Du, H., et al. 2019. Cellulose nanocrystals and cellulose nanofibrils based hydrogels for biomedical applications. *Carbohydrate Polymers* 209: 130–144.
6. Ma, T., et al. 2022. Nanocellulose: A promising green treasure from food wastes to available food materials. *Critical Reviews in Food Science and Nutrition* 62 (4): 989–1002.
7. Pinto, R.J., et al. 2020. Highly electroconductive nanopapers based on nanocellulose and copper nanowires: A new generation of flexible and sustainable electrical materials. *ACS Applied Materials & Interfaces* 12 (30): 34208–34216.
8. Tang, C., et al. 2019. Pickering emulsions stabilized by hydrophobically modified nanocellulose containing various structural characteristics. *Cellulose* 26: 7753–7767.
9. Ding, Z., X. Yang, and Y. Tang. 2023. Nanocellulose-based electrodes and separator toward sustainable and flexible all-solid-state supercapacitor. *International Journal of Biological Macromolecules* 228: 467–477.
10. Wang, Y., et al. 2022. Nanolignin filled conductive hydrogel with improved mechanical, anti-freezing, UV-shielding and transparent properties for strain sensing application. *International Journal of Biological Macromolecules* 205: 442–451.
11. Jia, C., et al. 2017. Using a fully recyclable dicarboxylic acid for producing dispersible and thermally stable cellulose nanomaterials from different cellulosic sources. *Cellulose* 24: 2483–2498.
12. Wang, R., et al. 2017. Tailored and integrated production of carboxylated cellulose nanocrystals (CNC) with nanofibrils (CNF) through maleic acid hydrolysis. *ChemNanoMat* 3 (5): 328–335.
13. Habibi, Y., L.A. Lucia, and O.J. Rojas. 2010. Cellulose nanocrystals: Chemistry, self assembly, and applications. *Chemical Reviews* 110 (6): 3479–3500.

14. Isogai, A., and L. Bergström. 2018. Preparation of cellulose nanofibers using green and sustainable chemistry. *Current Opinion in Green and Sustainable Chemistry* 12: 15–21.
15. Moon, R.J., et al. 2011. Cellulose nanomaterials review: Structure, properties and nanocomposites. *Chemical Society Reviews* 40 (7): 3941–3994.
16. Isogai, A., T. Saito, and H. Fukuzumi. 2011. TEMPO-oxidized cellulose nanofibers. *Nanoscale* 3(1): 71–85.
17. Saito, T., et al. 2007. Cellulose nanofibers prepared by TEMPO-mediated oxidation of native cellulose. *Biomacromolecules* 8 (8): 2485–2491.
18. Lin, C., et al. 2021. Carboxylated cellulose nanocrystals with chiral nematic property from cotton by dicarboxylic acid hydrolysis. *Carbohydrate Polymers* 264: 118039.
19. Reyes, G., et al. 2019. Dissolution and hydrolysis of bleached kraft pulp using ionic liquids. *Polymers* 11 (4): 673.
20. Wei, X., et al. 2023. Research on deep eutectic solvents for the construction of humidity-responsive cellulose nanocrystal composite films. *International Journal of Biological Macromolecules* 235: 123805.
21. Loow, Y.-L., et al. 2018. Deep eutectic solvent and inorganic salt pretreatment of lignocellulosic biomass for improving xylose recovery. *Bioresource Technology* 249: 818–825.
22. Tan, Y.T., A.S.M. Chua, and G.C. Ngoh. 2020. Deep eutectic solvent for lignocellulosic biomass fractionation and the subsequent conversion to bio-based products–a review. *Bioresource Technology* 297: 122522.
23. Zhu, Y., et al. 2023. Acidic and alkaline deep eutectic solvents (DESs) pretreatment of grapevine: Component analysis, characterization, lignin structural analysis, and antioxidant properties. *International Journal of Biological Macromolecules* 236: 123977.
24. Anastas, P., and N. Eghbali. 2010. Green chemistry: Principles and practice. *Chemical Society Reviews* 39 (1): 301–312.
25. Sirviö, J.A., M. Visanko, and H. Liimatainen. 2015. Deep eutectic solvent system based on choline chloride-urea as a pre-treatment for nanofibrillation of wood cellulose. *Green Chemistry* 17 (6): 3401–3406.
26. Laitinen, O., et al. 2017. Sustainable stabilization of oil in water emulsions by cellulose nanocrystals synthesized from deep eutectic solvents. *Cellulose* 24: 1679–1689.
27. Li, P., et al. 2017. Cellulose nanofibrils from nonderivatizing urea-based deep eutectic solvent pretreatments. *ACS Applied Materials & Interfaces* 9 (3): 2846–2855.
28. Chen, L., et al. 2019. A novel deep eutectic solvent from lignin-derived acids for improving the enzymatic digestibility of herbal residues from cellulose. *Cellulose* 26: 1947–1959.
29. Yu, W., et al. 2019. Choline chloride-based deep eutectic solvent systems as a pretreatment for nanofibrillation of ramie fibers. *Cellulose* 26: 3069–3082.
30. Sirvio, J.A., et al. 2014. Strong, self-standing oxygen barrier films from nanocelluloses modified with regioselective oxidative treatments. *ACS Applied Materials & Interfaces* 6 (16): 14384–14390.
31. Li, P., et al. 2018. Recyclable deep eutectic solvent for the production of cationic nanocelluloses. *Carbohydrate Polymers* 199: 219–227.
32. Suopajärvi, T., J.A. Sirviö, and H. Liimatainen. 2017. Nanofibrillation of deep eutectic solvent-treated paper and board cellulose pulps. *Carbohydrate Polymers* 169: 167–175.
33. Sirviö, J.A., J. Ukkola, and H. Liimatainen. 2019. Direct sulfation of cellulose fibers using a reactive deep eutectic solvent to produce highly charged cellulose nanofibers. *Cellulose* 26: 2303–2316.
34. Liu, Q., et al. 2019. Choline chloride-lactic acid deep eutectic solvent for delignification and nanocellulose production of moso bamboo. *Cellulose* 26: 9447–9462.
35. Lu, H., et al. 2018. A novel method to prepare lignocellulose nanofibrils directly from bamboo chips. *Cellulose* 25: 7043–7051.
36. Sirviö, J.A. 2019. Fabrication of regenerated cellulose nanoparticles by mechanical disintegration of cellulose after dissolution and regeneration from a deep eutectic solvent. *Journal of Materials Chemistry A* 7 (2): 755–763.

37. Selkälä, T., et al. 2016. Anionically stabilized cellulose nanofibrils through succinylation pretreatment in urea–lithium chloride deep eutectic solvent. *Chemsuschem* 9 (21): 3074–3083.
38. Sirviö, J.A. 2018. Cationization of lignocellulosic fibers with betaine in deep eutectic solvent: Facile route to charge stabilized cellulose and wood nanofibers. *Carbohydrate Polymers* 198: 34–40.
39. Liu, C., et al. 2020. Production of lignin-containing cellulose nanofibers using deep eutectic solvents for UV-absorbing polymer reinforcement. *Carbohydrate Polymers* 246: 116548.
40. Rojo, E., et al. 2015. Comprehensive elucidation of the effect of residual lignin on the physical, barrier, mechanical and surface properties of nanocellulose films. *Green Chemistry* 17 (3): 1853–1866.
41. Vandamme, E.J., S. Baets, and A. Steinbüchel. 2002. *Biopolymers: Polysaccharides II, Polysaccharides from Eukaryotes*. Wiley-Vch.
42. Medronho, B., and B. Lindman. 2014. Competing forces during cellulose dissolution: From solvents to mechanisms. *Current Opinion in Colloid & Interface Science* 19 (1): 32–40.
43. Domínguez de María, P. 2014. Recent trends in (ligno) cellulose dissolution using neoteric solvents: switchable, distillable and bio-based ionic liquids. *Journal of Chemical Technology & Biotechnology* 89(1): 11–18.
44. Heinze, T., and A. Koschella. 2005. Solvents applied in the field of cellulose chemistry: A mini review. *Polímeros* 15: 84–90.
45. Fang, L., et al. 2010. Experimental and theoretical evidence of enhanced ferromagnetism in sonochemical synthesized $BiFeO_3$ nanoparticles. *Applied Physics Letters* 97 (24): 242501.
46. Ren, H., et al. 2016. The properties of choline chloride-based deep eutectic solvents and their performance in the dissolution of cellulose. *BioResources* 11 (2): 5435–5451.
47. Liu, Y., et al. 2017. Efficient cleavage of lignin–carbohydrate complexes and ultrafast extraction of lignin oligomers from wood biomass by microwave-assisted treatment with deep eutectic solvent. *Chemsuschem* 10 (8): 1692–1700.
48. Wu, W., et al. 2022. Structural and rheological properties of nanocellulose with different polymorphs, nanocelluloses I and II, prepared by natural deep eutectic solvents from sugarcane bagasse. *International Journal of Biological Macromolecules* 220: 892–900.
49. Sirviö, J., M. Visanko, and H. Liimatainen. 2016. Acidic deep eutectic solvents as hydrolytic media for cellulose nanocrystal production. *Biomacromolecules* 17: 3025–3032.
50. Troter, D.Z., et al. 2016. Application of ionic liquids and deep eutectic solvents in biodiesel production: A review. *Renewable and Sustainable Energy Reviews* 61: 473–500.
51. Xia, Q., et al. 2018. Multiple hydrogen bond coordination in three-constituent deep eutectic solvents enhances lignin fractionation from biomass. *Green Chemistry* 20 (12): 2711–2721.
52. Liu, Y., et al. 2017. Efficient cleavage of strong hydrogen bonds in cotton by deep eutectic solvents and facile fabrication of cellulose nanocrystals in high yields. *ACS Sustainable Chemistry & Engineering* 5 (9): 7623–7631.
53. Wang, H., et al. 2020. Extraction of cellulose nanocrystals using a recyclable deep eutectic solvent. *Cellulose* 27: 1301–1314.
54. Yang, X., et al. 2019. Facile extraction of thermally stable and dispersible cellulose nanocrystals with high yield via a green and recyclable $FeCl_3$-catalyzed deep eutectic solvent system. *ACS Sustainable Chemistry & Engineering* 7 (7): 7200–7208.
55. Pérez-Pérez, A., et al. 2023. Synergetic effect of hydrothermal and deep eutectic solvents (DES) pretreatments on Robinia wood fractionation for the manufacture of bioethanol and cellulose nanocrystals. *Industrial Crops and Products* 203: 117157.
56. Zhang, H., et al. 2022. Separation cellulose nanocrystals from microcrystalline cellulose using hydrated deep eutectic solvent and high shear force. *Industrial Crops and Products* 189: 115781.
57. Douard, L., et al. 2021. Natural acidic deep eutectic solvent to obtain cellulose nanocrystals using the design of experience approach. *Carbohydrate Polymers* 252: 117136.
58. Sirvio, J.A., M. Visanko, and H. Liimatainen. 2016. Acidic deep eutectic solvents as hydrolytic media for cellulose nanocrystal production. *Biomacromolecules* 17 (9): 3025–3032.
59. Jiang, J., et al. 2020. High production yield and more thermally stable lignin-containing cellulose nanocrystals isolated using a ternary acidic deep eutectic solvent. *ACS Sustainable Chemistry & Engineering* 8 (18): 7182–7191.

60. Ling, Z., et al. 2019. Structural variations of cotton cellulose nanocrystals from deep eutectic solvent treatment: Micro and nano scale. *Cellulose* 26 (2): 861–876.
61. Liu, S., et al. 2021. Esterification of cellulose using carboxylic acid-based deep eutectic solvents to produce high-yield cellulose nanofibers. *Carbohydrate Polymers* 251: 117018.
62. Wu, M., et al. 2021. Integrated and sustainable preparation of functional nanocellulose via formic acid/choline chloride solvents pretreatment. *Cellulose* 28 (15): 9689–9703.
63. Lim, W.-L., et al. 2021. Cellulose nanocrystals from bleached rice straw pulp: Acidic deep eutectic solvent versus sulphuric acid hydrolyses. *Cellulose* 28 (10): 6183–6199.
64. Smirnov, M.A., et al. 2020. Green method for preparation of cellulose nanocrystals using deep eutectic solvent. *Cellulose* 27: 4305–4317.
65. Ling, Z., et al. 2021. Promoting enzymatic hydrolysis of aggregated bamboo crystalline cellulose by fast microwave-assisted dicarboxylic acid deep eutectic solvents pretreatments. *Bioresource Technology* 333: 125122.
66. Suopajärvi, T., et al. 2020. Acidic and alkaline deep eutectic solvents in delignification and nanofibrillation of corn stalk, wheat straw, and rapeseed stem residues. *Industrial Crops and Products* 145: 111956.
67. Wang, S., et al. 2023. A novel one-pot strategy to construct 3D-printable cellulose nanofiber/poly (deep eutectic solvent) conductive elastomers. *Chemical Engineering Journal* 454: 140022.
68. Jiang, J., et al. 2020. Acidic deep eutectic solvent assisted isolation of lignin containing nanocellulose from thermomechanical pulp. *Carbohydrate polymers* 247: 116727.
69. Jiang, J., et al. 2021. Rapid, high-yield production of lignin-containing cellulose nanocrystals using recyclable oxalic acid dihydrate. *Industrial Crops and Products* 173: 114148.

Chapter 6
Nanocellulose in Textile Applications

Abstract With the boom of nanotechnology, textile industries are extensively getting development opportunities. Textile product quality is being enhanced for specific and generally used applications with nanomaterials embedded into the polymer matrix part coated onto the fiber surface. The sources of nanocellulose (NC) are very common in nature and these NCs are generally extracted from the cell walls of plants. Nano cellulose are the materials showing diversity in their properties including excellent stiffness, good strength, and higher surface area values. Textile material surface modification is possibility done with the higher number of hydroxyl groups presented in nanocellulose chemical structure. Additionally, nanocellulose has been reported as eco-friendly and is not sensitive to environmental changes. This section of the study reveals the effects and applications of nanocellulose on textile modification.

6.1 Introduction

Textile materials have always played a crucial role in human life, generally prepared from natural fibers offering essential properties to textile products that contribute to high-quality materials. Cellulose an abundant material on earth is widely utilized for its unique characteristics and has been derived from vegetable biomass, cotton, and wood. Among these fibers, cellulose stands out due to its availability and versatility. The powder form of cellulose is generally known as nanocellulose or microfibrillated cellulose, which is being used in different applications. Nanocellulose materials are also recognized as renewable, biodegradable, and biocompatible materials which are not only used in textile industrial applications, but, they have also been used in biomedical and food packaging industries [1, 2].

Generally, natural fibers with small principal lengths are not suitable for spinning yarns due to higher variations in their lengths. As a result, cotton, wool, hemp, or silk fibers in small lengths during the processing stages and finishing usages in textile manufacturing are often collected as waste. Thus, a novel approach is being practiced to recycle these fibers by converting these types of waste into nanocellulose.

A. Farooq et al., *Deep Eutectic Solvents in the Textile Industry*,
https://doi.org/10.1007/978-981-97-6433-4_6

Nanocellulose derived from sources such as cotton, wood pulp, or agricultural waste, possesses superior properties that make it highly desirable in various industries. Its versatility, sustainability, market potential, and increased value contribute to its appeal as a valuable material for applications in textiles, composites, coatings, and biomedical fields [1, 3, 4].

Over the last ten years, there has been a significant increase in interest in using nanocellulose, primarily driven by its unique chemical and physical properties. Nanocellulose can be broadly categorized into two main types: (i) cellulose nanocrystals (CNC), obtained through acid treatment, and (ii) cellulose nanofibrils (CNF), primarily produced via mechanical disintegration [5]. Cellulose materials are widely used as nanofillers in composite materials, they are often subjected to hydrolysis by using strong acids including sulfuric acid (H2SO4) or hydrochloric acid (HCl). This hydrolysis process selectively breaks down the amorphous regions of cellulose, leading to the disintegration of microfibril. As a result, crystalline nanofibers, commonly known as nanowhiskers, are formed [1]. Similarly, cellulose nanofibrils are also being used in various applications readily available in the market as a commercial product which is increasingly gaining interest in various industrial applications as evident from the recent studies [6] and book chapters [7–9] that highlight its potential. The common methods used in cellulose nanofibrils are 2,2,6,6-tetramethylpiperidine-N-oxyl (TEMPO)-mediated oxidation [3] or enzymatic hydrolysis [10–12], which have facilitated the mechanical disintegration process and made CNF a more attractive material for commercial applications. Thus, this chapter focuses on the impact of nanocellulose on textile properties as well as applications of nanocellulose in textile materials production.

6.2 Nanocellulose in Textile Applications

In recent decades, there has been a tremendous shift in the utilization and application of natural and renewable resources, driven by growing concerns surrounding environmental conservation. As a result, nanocellulose has emerged as a material of considerable interest in various industries, including the medical, automotive, and textile sectors [13]. This extensive increase in attention can be attributed to the exceptional properties exhibited by nanocellulose, such as its small dimensions, high mechanical strength, significant specific surface area, biodegradability, and biocompatibility. These distinctive features have gained significant attention from researchers, positioning nanocellulose as a highly promising and versatile material with substantial potential across diverse applications [2, 14]. Nanocellulose is an extraordinary material that has garnered considerable attention due to its elegant and remarkable properties, particularly in terms of its applications in nano-products. It is a form of cellulose that exists at the nanometric scale, with dimensions smaller than 100 nm. Nanocellulose has emerged as a revolutionary material with remarkable features, particularly in the textile industry. The application of nanotechnology, specifically nanocellulose, has the potential to further enhance and transform the fibers used in textile production. The combination of nanocellulose's advantageous properties, such as low cost, environmentally friendly nature, light weight, high resistance, and electric conductivity, has paved the way for a new generation of textiles and opened up a wide range of possible applications [15].

Nanocellulose is a class of materials that can be categorized into three distinct types: (I) cellulose nanocrystals (CNCs), also known as nanocrystalline cellulose (NCC) and cellulose nanowhiskers (CNWs), (II) cellulose nanofibrils (CNFs), also referred as nano-fibrillated cellulose (NFC), and (III) bacterial cellulose (BC). Various extraction methods are being employed to isolate nanoparticles from cellulose sources, leading to particles with diverse characteristics such as crystallinity, surface chemistry, and mechanical properties [16].

Currently, cellulose nanocrystals are primarily produced using acid hydrolysis/heat-controlled techniques, with sulfuric acid being the most commonly employed acid. The process involves selectively hydrolyzing the amorphous regions of cellulose fibers, resulting in the formation of highly crystalline particles with dimensions that depend on the source material. For example, plant-derived CNCs typically have dimensions of 5–20 nm × 100–500 nm. Sulfuric acid hydrolysis introduces negatively charged sulfate half-ester groups onto the particle surfaces, which facilitate the prevention of aggregation in aqueous suspensions through electrostatic repulsion. On the other hand, cellulose nanofibrils are entangled fibrils that have micrometer-scale lengths and contain both amorphous and crystalline cellulose regions. Unlike CNCs, CNFs exhibit a slightly lower degree of crystallinity (approximately 90%) but possess entangled structures. Consequently, aqueous suspensions of CNFs at relatively low concentrations (below 1 wt.%) exhibit high viscosity. The extraction of CNFs from cellulosic fibers can be accomplished through three types of processes: (I) mechanical treatments such as homogenization, grinding, and milling, (II) chemical treatments like TEMPO oxidation, and (III) a combination of chemical and mechanical treatments [17]. In the context of textile industries, nanocellulose has emerged as a promising material, offering numerous outstanding applications. Some significant nanocellulose applications in textile industries are included as follows.

6.3 Influence of Nanocellulose on Textile Properties

In a study focused on investigating the effects of nanocellulose on textile properties, the study specifically examined the physical properties of polyester fabrics [18]. The findings revealed notable changes in these properties, suggesting the potential of nanocellulose to positively impact the performance of textiles. The study involved the evaluation of both treated and untreated polyester fabrics, with a focus on assessing the changes in their physical properties, particularly in terms of breaking load and crease recovery angle.

6.3.1 Effect on Tensile Strength

When the pristine and treated polyester fabrics were evaluated for tensile strength properties the results exhibited that the load-bearing capacity of the fibers was increased, because of the penetration and diffusion of nanocellulose into the polymer matrix of fabric samples [19]. Thus the tensile strength properties of the polyester fabrics were improved.

6.3.2 Effect of Nanocellulose on Water Absorbency Performance

The effect of nanocellulose on polyester fabric's water absorbency performance was measured by using a drop test and a wicking behavior test [20, 21]. The results of the study revealed that water absorbency of fabric samples exhibits less time for absorption of water when the fabrics were treated with 1 g per liter (gpl) concentration of nanocellulose. However, the time for absorption of water was increased when the nanocellulose concentration was increased from 5 to 10 gpl, which shows a reduction in the absorbency of textile material when coated with nanocellulose. The hydrophilicity behavior exhibited by the polyester fabric sample was due to a lower concentration of nanocellulose. When the concentration of NC in the coating of polyester fabric was increased, it was observed that more time was required for the absorption of water droplets. This is attributed to the blocking effect of NC, which hinders the diffusion of water molecules and restricts their accommodation within the spaces between polymer chains, as a result.

6.3.3 Effect of Nanocellulose on the Air Permeability of Polyester Fabric

The effect of nanocellulose on the air permeability of polyester fabrics was done by comparing the untreated fabric samples with the fabrics coated with various concentrations of nanocellulose. The results highlight that fabric samples treated with nanocellulose have low air permeability as compared to untreated samples, possibly because of the hindering of nanocellulose in polymer matric towards the flow of air through the fabric sample.

6.3.4 *Effect of Nanocellulose on Water Permeability of Polyester Fabric*

The results demonstrated a wide reduction in water permeability through the polyester fabric sample that has been treated with nanocellulose compared to the untreated sample. This reduction can be attributed to the resistance existing by the nanoparticles present in the polymer matrix, which block the flow of water through the fabric. Furthermore, as the concentration of nanoparticles in the fabric increases, the permeability of water further decreases. This observation highlights the potential of nanocellulose treatment in enhancing the water resistance properties of polyester fabric, exhibiting the potential benefits in various applications where water repellency is desired.

6.3.5 *Effect of Nanocellulose on Dyeing of Treated Fabric with Direct Dye*

Nano cellulose treatment excellently increases the color strength of polyester fabric samples dyed with direct dyes, additionally, results revealed the excellent improvement in fastness towards fabric soaping. The results obtained from the experiment indicate that as the amount of dye increases, there is a corresponding increase in the K/S value of the sample treated with a direct dye. The dyed polyester fiber samples, to which nanocellulose was added, were subjected to a washing process using nonionic detergent at a temperature of 70 °C for 15 min. It was observed that the decrease in K/S values was more pronounced in the samples dyed without the addition of nanocellulose. This suggests that the presence of nanocellulose provides a protective effect, resulting in a lesser reduction of the dye's color strength (K/S value) during the washing process compared to samples without nanocellulose treatment.

6.3.6 *Functionalization of Nanocellulose in the Textile Industry*

The functionalization and modification of nanocellulose's hydroxyl groups present a significant potential for expanding its application potential in diverse fields, especially in textile industries. Through simple reactions like acetylation and oxidation processes, the surface of nanocellulose can be modified with various chemical functionalities [22]. Additionally, functionalization sometimes involves grafting functional elements and polymers onto the surface of nanocellulose. The goals of functionalization consist of several aspects, including introducing functional groups tailored for specific applications [23], altering the affinity of nanocellulose towards polar and nonpolar molecules [24], and modulating the surface characteristics of nanocellulose to enhance its compatibility with nonpolar matrices, thereby improving processing capabilities [25].

6.4 Nanocellulose Coating in Textile Dyeing and Ultraviolet Protection

The textile industry has a keen interest in utilizing nitrocellulose-based coatings not only to improve the mechanical properties but also to address crucial challenges associated with different stages of textile materials production. These challenges include reducing the generation of hazardous wastage, particularly during coating application and textile dyeing processes. One complexity in textile applications stems from the fact that not all fabrics are made of cellulose. As a result, the requirements for chemical functionalization of nanocellulose may vary depending on the type of fiber, whether natural or synthetic, that requires coating. This is because the surface functionalization of nanocellulose needs to ensure proper adhesion to the substrate. Thus in a recent study [18], a substrate made of 100% polyester was subjected to treatment using nanocrystal suspensions derived from scraps of viscose rayon fibers. Following the treatment, the substrate was dyed using a direct dye (Congo red BDC). The results obtained confirm the correlation between the size and shape of the synthesized nanocellulose particles. The dyeing behavior of the treated fabrics with direct dye was also investigated, and the buildup of dyes was measured in terms of color strength, represented by K/S values. It was observed that higher K/S values were achieved when cellulose nanoparticles were anchored within the fiber matrix, particularly after pretreatment and dyeing with direct dyes. Moreover, the treatment of fabrics with nanocellulose resulted in improved color strength and exhibited good resistance to soaping.

Nanocellulose coating offers the inherent ability to enhance the mechanical properties of coated textiles. This advantageous characteristic, combined with the biocompatibility and safety of nanocellulose, opens up possibilities for utilizing cellulose nanofibril coatings in intelligent textiles used for medical applications, such as lumbar belts [26]. These protective garments are designed to prevent and mitigate the risks of spinal injuries resulting from excessive workload or inadequate care by workers during their tasks. An ideal intelligent fabric for such applications should exhibit adaptive behavior based on various usage conditions, including pressure, load, humidity, temperature, and more. To achieve this, poly(ethylene terephthalate) (PET) samples were coated with CNFs and potato starch gel. The inclusion of starch aimed to improve the adhesion of the CNF coating to the polyester substrate. It is important to note that the water-soluble nature of the coating, primarily due to the solubility of starch, renders it unsuitable for protective garments. However, traction tests indicated a significant increase in the elasticity of the textile when impregnated with nanocellulose and starch. Consequently, the CNF coating presents an attractive

Fig. 6.1 Photographs of 100% cotton woven fabric samples before and after five accelerated laundry cycle washing colored with cellulose nanofibrils dyes **a** blue, **b** black, **c** red, and **d** yellow [27]

method for enhancing the mechanical properties of intelligent fabrics, provided an effective anchoring technique can be devised for synthetic fibers such as PET.

In a recent study, environmental pollution by effluents discharged from current textile dyeing processes has been aimed to minimize by using cellulose nanofibrils [27]. The authors have demonstrated a post-treatment method for CNF-colored cotton textiles using polycarboxylic acid. This post-treatment ensures the permanent chemical grafting of the soluble dye-labeled poly-sugars and forms chemical cross-links with the CNF fibers present in the cotton fabric through esterification reactions. As a result, dye fixation is improved by 30%, and the discharge of dyes during the washing stage is reduced by 60%. These enhancements were excellently achieved without compromising the stiffness and breathability of the fabrics as shown in Fig. 6.1. The effectiveness of this advanced textile method was tested using a range of reactive dyes that cover the entire visual spectrum range.

Coating textiles with ultraviolet (UV) protection properties can play a crucial role in preservation individuals against the diffusion of skin cancer. Such coatings provide textile garments that effectively shield the skin from the harmful effects of sunburn. Typically, UV-protective coatings consist of a transparent layer containing materials that absorb UV light. This UV-absorbing layer can be either organic or inorganic. The UV-absorbing materials need to exhibit strong absorption (high extinction coefficient) within the UV range of 290–360 nm. The effectiveness of a protective garment is usually evaluated using parameters such as the UV protection factor (UPF) and the sun protection factor (SPF) [28]. Nanostructures and natural materials are considered highly effective in providing UV protection coatings. Nanomaterials exhibit high activity in this regard, while natural materials are particularly beneficial for human health and the environment.

To investigate the possibility of preparing a UV-blocking layer for cotton fabrics, researchers tested three different types of cellulose nanocrystals that had distinct particle sizes. The goal was to evaluate the effectiveness of these CNCs in providing UV protection for the fabric [29]. Generally, Cellulose nanocrystals, when combined with UV-blocking additives, can be utilized to create an anti-UV layer. This combination has yielded an effective UV-protective coating for textiles in this study, mitigating the harmful effects of UV radiation on the skin. A pad-dry-cure process was employed to modify cotton fabric with a biocompatible and non-toxic nanocomposite comprised of chitosan (CS) and nanocellulose (NCs). This modification aimed to enhance the properties of the cotton fabric while ensuring biocompatibility and safety. To understand the impact of nanocellulose particle size on the morphology and properties of cotton fabrics modified with chitosan (CS)/NCs nanocomposites, a comprehensive analysis was conducted using scanning electron microscopy (SEM), Fourier transforms infrared spectroscopy (FTIR), mechanical property testing, water wettability assessment, UV protection analysis, and laundering durability evaluation was done. The results revealed that the CS/NCs-coated fabrics exhibited excellent UV protection capabilities and maintained good laundering durability. The fabrics coated with CS/NCs displayed significantly higher UPF values compared to the untreated fabric, even after undergoing 30 cycles of laundering. However, it was observed that the particle size of the NCs harmed the UV protection ability of the CS/NCs-coated fabric. The coated fabric samples exhibited improved strength and durability, while also displaying increased resistance to water penetration.

6.5 Nanocellulose in Medical Tools

Nanocellulose possesses valuable properties such as compatibility with the human body, ability to break down naturally, a large surface area, biocompatible, biodegradable, low cost, and impressive strength [2]. Thus, nanocellulose in fabric development has the potential to bring significant benefits to the medical sector, particularly in wound treatment applications [30]. Bacterial nanocellulose (BNC) has gained recognition for its exceptional properties that make it highly suitable for biomedical applications among all types of nanocellulose mentioned earlier including wound healing and tissue engineering [31, 32]. Additionally, several bacterial nanocellulose-based wound dressing products are commercially available in the market such as Bioprocess, XCell, and Biofill [32]. Because of the excellent numerous advantages of nanocellulose researchers have used it in biomedical applications for the development of wound dressings. Thus in a study [33], authors have investigated the potential of nanocellulose as a bioink for modifying film surfaces using a bioprinting technique. Two different types of nanocellulose were prepared and utilized, one prepared through TEMPO-mediated oxidation and the other through a combination of carboxymethylation and periodate oxidation. The results of the study show that the reduced dimensions of nanofibrils resulted in lower viscosity of the nanocellulose, providing favorable rheological properties for its use as a bioink. Additionally, results observed from this study exhibited that both types of nanocellulose showed an interesting resistance to bacterial growth, further highlighting the potential of these novel materials. The researchers utilized Gluconacetobacter xylinus to create bacterial nanocellulose from a 2% glucose solution [31]. The BNC was then evaluated for its effects on cells (cyto-evaluation). The findings revealed that NIH/3T3 cells exhibited strong compatibility and minimal toxicity when exposed to BNC. Based on these results, the BNC produced in this investigation was recommended as a promising option for wound dressings in extensive skin transplants and reconstructive surgeries. The progression of burn wounds poses a notable challenge, as burns initially perceived as superficial can evolve into full-thickness burns over time. Cooling has proven to be an effective approach in mitigating the conversion of burn wounds. However, if the cooling agent is colder than room temperature, there is a risk of hypothermia for the patient, particularly when considering the size of the wound. Moreover, this low temperature can impede tissue perfusion, exacerbating the progression of the burn wound. To address the issue, a study reported aimed to investigate whether wound dressings made from non-pre-cooled bacterial nanocellulose, with a high water content, could cool burns through evaporation alone and reduce intradermal damage in the skin [34].

6.6 Nanocellulose in the Military Sector

Textiles play a vital role in military equipment, serving diverse functions across a range of applications. In the military context, textiles find application in various items such as combat uniforms, tents, shelters, parachutes, sails, airplane materials, safety harnesses, body armor, bomb suits for explosive ordnance disposal, sandbags, fire-resistant materials, footwear, camouflage systems, and many other products. These textiles contribute significantly to the functionality and effectiveness of military operations [35, 36].

The textile industry is actively exploring the integration of technology to develop advanced garments with scientific properties, such as antimicrobial properties, fire resistance, self-cleaning abilities, and improved water and air permeability. These advancements in textile products have found applications in both commercial and military sectors, albeit with some variations in their specific objectives. In military contexts, the focus is primarily on the development of hydrophobic textiles, that possess high strength, and exhibit fire-retardant properties. Nanocellulose materials known for their highly adaptable properties, are considered a promising option for integration into military products, such as uniforms. The National Research Council has emphasized that advanced materials used in the military should meet a wide range of criteria, including cost-effectiveness, strength, durability, speed, eco-friendliness, accuracy, low weight, and compatibility [34, 37].

Nanocellulose has found applications in modern textiles, offering a range of benefits because of its excellent properties [38]. Its superior properties compared to materials like Kevlar make it a valuable resource for various purposes, including thermal insulation, sanitary napkins, tampons, diapers, and wound dressings. Innovations in this field have led to the identification of numerous applications for nanocellulose in the textile industry. Mishra et al. [39] discussed in their review [39] the potential

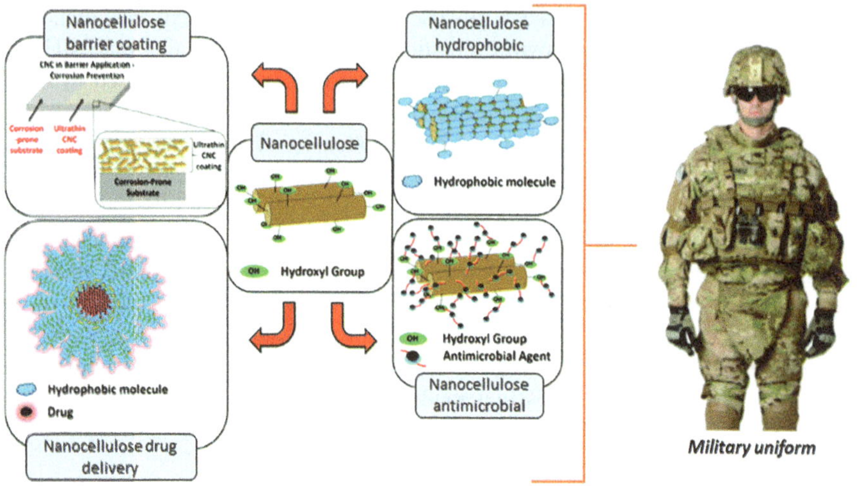

Fig. 6.2 Nanocellulose in military uniform, figure redrawn and adapted from [40, 41]

of nanocellulose in producing lightweight body armor and ballistic glass. Figure 6.2 illustrates the specific characteristics required for military-grade body armor, such as high mechanical strength, lightweight, hydrophobicity, antimicrobial properties, and anti-ballistic capabilities. Nanocellulose emerges as a promising alternative to the current material, Kevlar, for manufacturing body armor.

The process of selecting materials for military equipment fabrication is a complex task. It requires careful consideration to ensure that the chosen components provide comprehensive advantages for soldiers. In this regard, the potential applications of nanocellulose have been extensively examined, particularly in meeting the requirements for lightweight body armor and ballistic glass [39]. Significant endeavors have been made to incorporate nanocellulose technology into military appliances. Superhydrophobic nanocellulose-based textiles hold significant potential for a wide range of applications in military settings. During military operations, soldiers often encounter wet and dirty conditions, necessitating the need for clothing and materials that can effectively repel water and remove contaminants. To achieve superhydrophobicity in nanocellulose-based textiles, two key factors highlighted by Wei et al. [42] are crucial (1) the creation of a suitable hierarchical structure incorporating durable nanoparticles, and (2) the development of a low-energy surface. Various methods have been employed to fabricate robust superhydrophobic textile surfaces, including dip-coating, wet chemical deposition, electro-assisted chemical deposition, spray coating, sol–gel techniques, chemical etching, chemical vapor deposition, plasma processing, and polymer grafting.

The utilization of electronic textiles in military applications has been extensively examined by [43]. These electronic textiles offer various functionalities, including personnel health monitoring (such as blood pressure, heartbeat, and body temperature, soldier tracking, communication capabilities, and environmental temperature monitoring. In this regard, nanocellulose functionalization has proven successful in enhancing the conductivity of these textile materials [44]. In military contexts, electronic textiles often employ electrical impulses as output signals. For instance, acoustic materials are being used to detect approaching unidentified vehicles, enemies, or aircraft, while buttons or optical detectors utilizing infrared signals can detect gunshot locations [43]. Figure 6.3 illustrates the application of modified nanocellulose in electronic textiles for military purposes. Nechyphorchuk et al. [45] developed woven cotton fabrics incorporating cellulose nanofibrils (CNF) mixed with plasticizers, suitable for printing or dyeing using conventional colored water-based inks or conductive silver nanoparticle inks. The presence of CNF facilitated ink settling onto the substrate, enabling high-quality printing at faster speeds. Additionally, the CNF coating provided a conductive layer that maintained the electric signal even when the fabric was folded. Various fabrication methods are commonly employed in the production of nanocellulose-based textiles. These approaches include chemical etching, chemical vapor deposition, dip-coating, electro-assisted chemical deposition, spray coating, sol–gel techniques, plasma processing, and polymer grafting, as well as wet chemical deposition. These methods offer different advantages and can be selected based on specific requirements and desired properties of the resulting textiles.

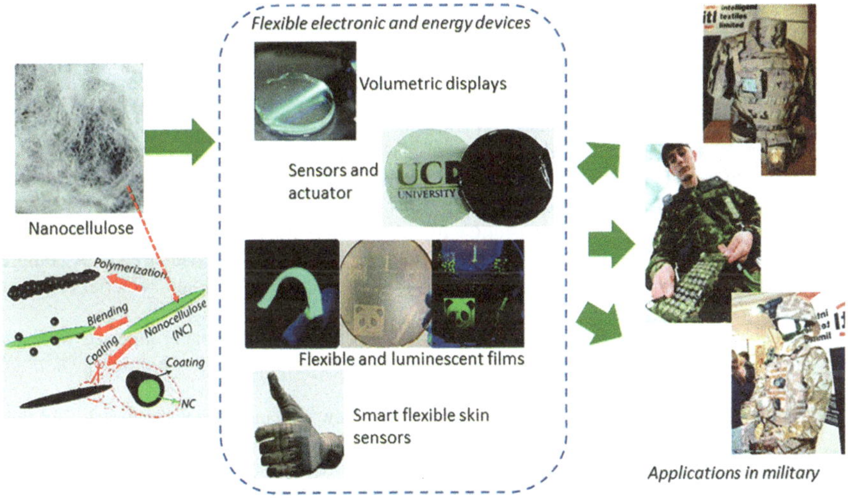

Fig. 6.3 The integration of wearable computers and keyboards into military uniforms [46]

6.7 Conclusions

The increasing interest in low-cost and natural bioproducts has led researchers to focus on nanocellulose as a versatile biomaterial with applications in various industries. The textile industry, in particular, has recognized the advantageous properties of nanocellulose and its potential for diverse applications. This chapter has emphasized the significant potential of incorporating nanocellulose into the textile industry and has highlighted various functionalized nanocellulose materials that play a crucial role in producing advanced biomaterials for this sector. The chapter has also discussed the application of nanocellulose in textile-related industries, including biomedical tools and the military sector. Although there are challenges associated with nanocellulose technology, further intensive efforts are required to foster novel inventions and discoveries in this field.

References

1. Farooq, A., et al. 2020. Cellulose from sources to nanocellulose and an overview of synthesis and properties of nanocellulose/zinc oxide nanocomposite materials. *International Journal of Biological Macromolecules* 154: 1050–1073.
2. Farooq, A., et al. 2022. Extraction of cellulose nanofibrils from Ficus natalensis barkcloth and utilization in preparation of antimicrobial bio-nanocomposite films for possible food packaging applications. *Journal of Industrial Textiles*, 51(3_suppl): 3980S-3997S.
3. Farooq, A., et al. 2021. Structure and properties of high quality natural cellulose nano fibrils from a novel material Ficus natalensis barkcloth. *Journal of Industrial Textiles* 51 (4): 664–680.
4. Farooq, A., et al. 2022. Novel pretreatment performance evaluation for cellulose nanofibrils extraction from Ficus natalensis barkcloth. *Journal of Polymers and the Environment*, 1–13.
5. Nechyporchuk, O., M.N. Belgacem, and J. Bras. 2016. Production of cellulose nanofibrils: A review of recent advances. *Industrial Crops and Products* 93: 2–25.
6. Charreau, H., M.L. Foresti, and A. Vazquez. 2013. Nanocellulose patents trends: A comprehensive review on patents on cellulose nanocrystals, microfibrillated and bacterial cellulose. *Recent Patents on Nanotechnology*, 7(1): 56–80.
7. Bardet, R., and J. Bras. 2014. Cellulose nanofibers and their use in paper industry. In *Handbook of Green Materials: 1 Bionanomaterials: Separation Processes, Characterization and Properties*, 207–232. World Scientific.
8. Chauve, G., and J. Bras. 2014. Industrial point of view of nanocellulose materials and their possible applications. In *Handbook of Green Materials: 1 Bionanomaterials: Separation Processes, Characterization and Properties*, 233–252. World Scientific.
9. Lindström, T., et al. 2002. Microfibrillated cellulose. *Encyclopedia of Polymer Science and Technology*, 1–34.
10. Janardhnan, S., and M.M. Sain. 2006. Isolation of cellulose microfibrils–an enzymatic approach. *BioResources* 1 (2): 176–188.
11. Henriksson, M., et al. 2007. An environmentally friendly method for enzyme-assisted preparation of microfibrillated cellulose (MFC) nanofibers. *European Polymer Journal* 43 (8): 3434–3441.
12. Pääkkö, M., et al. 2007. Enzymatic hydrolysis combined with mechanical shearing and high-pressure homogenization for nanoscale cellulose fibrils and strong gels. *Biomacromolecules* 8 (6): 1934–1941.

13. Farooq, A., et al. 2022. An eco-friendly route to prepare cellulose based multifunctional lyocell fabrics using zinc oxide and cellulose nanofibrils network. *Fibers and Polymers* 23 (5): 1275–1283.
14. Jiang, S., et al. 2021. Investigation of a widely applicable process for extracting carboxyl-rich cellulose nanocrystal (CNC). *Fibers and Polymers* 22: 647–657.
15. Farooq, A., et al. 2020. Preparation and characterization of lyocell fabrics coated with cellulose nano-crystals and chitosan as functional materials. *Technical Journal* 25 (4): 21–25.
16. Klemm, D., et al. 2011. Nanocelluloses: A new family of nature-based materials. *Angewandte Chemie International Edition* 50 (24): 5438–5466.
17. Habibi, Y., H. Chanzy, and M.R. Vignon. 2006. TEMPO-mediated surface oxidation of cellulose whiskers. *Cellulose* 13: 679–687.
18. Chattopadhyay, D., and B. Patel. 2016. Synthesis, characterization and application of nano cellulose for enhanced performance of textiles. *Journal of Textile Science & Engineering* 6 (248): 2.
19. Zhong, T., et al. 2020. Nanocellulose from recycled indigo-dyed denim fabric and its application in composite films. *Carbohydrate Polymers* 240: 116283.
20. Chen, Q., et al. 2022. Wicking and drying behavior of knitted cotton fabrics under modified water drop test. *Journal of Natural Fibers* 19 (1): 232–247.
21. Tang, K.-P.M., C.-W. Kan, and J.-T. Fan. 2014. Evaluation of water absorption and transport property of fabrics. *Textile Progress* 46 (1): 1–132.
22. Wu, Z., et al. 2018. Preparation, characterization and acetylation of cellulose nanocrystal allomorphs. *Cellulose* 25: 4905–4918.
23. Morandi, G., and W. Thielemans. 2012. Synthesis of cellulose nanocrystals bearing photocleavable grafts by ATRP. *Polymer Chemistry* 3 (6): 1402–1407.
24. Laitinen, O., et al. 2016. Alkyl aminated nanocelluloses in selective flotation of aluminium oxide and quartz. *Chemical Engineering Science* 144: 260–266.
25. Salajková, M., L.A. Berglund, and Q. Zhou. 2012. Hydrophobic cellulose nanocrystals modified with quaternary ammonium salts. *Journal of Materials Chemistry* 22 (37): 19798–19805.
26. Spagnuolo, L., R. D'Orsi, and A. Operamolla. 2022. Nanocellulose for paper and textile coating: The importance of surface chemistry. *ChemPlusChem* 87 (8): e202200204.
27. Liyanapathiranage, A., et al. 2020. Nanocellulose-based sustainable dyeing of cotton textiles with minimized water pollution. *ACS Omega* 5 (16): 9196–9203.
28. Downs, N., and S. Harrison. 2018. A comprehensive approach to evaluating and classifying sun-protective clothing. *British Journal of Dermatology* 178 (4): 958–964.
29. Yang, X., et al. 2020. A biocompatible and sustainable anti-ultraviolet functionalization of cotton fabric with nanocellulose and chitosan nanocomposites. *Fibers and Polymers* 21: 2521–2529.
30. Puertas-Bartolomé, M., A. Mora-Boza, and L. García-Fernández. 2021. Emerging biofabrication techniques: A review on natural polymers for biomedical applications. *Polymers* 13 (8): 1209.
31. Fu, L., J. Zhang, and G. Yang. 2013. Present status and applications of bacterial cellulose-based materials for skin tissue repair. *Carbohydrate Polymers* 92 (2): 1432–1442.
32. Petersen, N., and P. Gatenholm. 2011. Bacterial cellulose-based materials and medical devices: Current state and perspectives. *Applied Microbiology and Biotechnology* 91: 1277–1286.
33. Rees, A., et al. 2015. 3D bioprinting of carboxymethylated-periodate oxidized nanocellulose constructs for wound dressing applications. *BioMed Research International*, 2015.
34. Holzer, J.C., et al. 2020. A novel human ex-vivo burn model and the local cooling effect of a bacterial nanocellulose-based wound dressing. *Burns* 46 (8): 1924–1932.
35. Steffens, F., et al. 2019. Military textiles-an overview of new developments. *Key Engineering Materials* 812: 120–126.
36. Ngô, C., and M. Van de Voorde. 2014. Nanotechnology in a Nutshell. *Nanotechnol a Nutshell*.
37. Luo, F., et al. 2017. A novel intumescent flame retardant with nanocellulose as charring agent and its flame retardancy in polyurethane foam. *Polymer Composites* 38 (12): 2762–2770.
38. Yetisen, A.K., et al. 2016. Nanotechnology in textiles. *ACS Nano* 10 (3): 3042–3068.

39. Mishra, R., et al. 2014. The production, characterization and applications of nanoparticles in the textile industry. *Textile Progress* 46 (2): 133–226.
40. Krueger, G. 2012. Psychological issues in military uniform design. In *Advances in Military Textiles and Personal Equipment*, 64–82e. Elsevier.
41. Panchal, P., E. Ogunsona, and T. Mekonnen. 2018. Trends in advanced functional material applications of nanocellulose. *Processes* 7 (1): 10.
42. Wei, D.W., et al. 2020. Superhydrophobic modification of cellulose and cotton textiles: Methodologies and applications. *Journal of Bioresources and Bioproducts* 5 (1): 1–15.
43. Textiles, E. 2015. *SMART FABRICS AND WEARABLE TECHNOLOGY*, ed. Dias, T. Woodhead Publishing.
44. Hebeish, A., et al. 2016. Advancement in conductive cotton fabrics through in situ polymerization of polypyrrole-nanocellulose composites. *Carbohydrate Polymers* 151: 96–102.
45. Nechyporchuk, O., et al. 2017. Cellulose nanofibril-based coatings of woven cotton fabrics for improved inkjet printing with a potential in e-textile manufacturing. *ACS Sustainable Chemistry & Engineering* 5 (6): 4793–4801.
46. Norrahim, M.N.F., et al. 2021. Nanocellulose: The next super versatile material for the military. *Materials Advances* 2 (5): 1485–1506.

The manufacturer's authorised representative in the EU is Springer Nature Customer Service Centre GmbH, Europaplatz 3, 69115 Heidelberg, Germany. If you have any concerns regarding our products, please contact ProductSafety@springernature.com

Printed and bound by CPI Group (UK) Ltd, Croydon, CR0 4YY
01/10/2026
02225875-0002